HIGH-FREQUENCY SWITCHING POWER SUPPLIES

McGraw-Hill Reference Books of Interest

Handbooks

Avallone and Baumeister • STANDARD HANDBOOK FOR MECHANICAL ENGINEERS
Benson • AUDIO ENGINEERING HANDBOOK
Benson • TELEVISION ENGINEERING HANDBOOK
Coombs • PRINTED CIRCUITS HANDBOOK
Coombs • BASIC ELECTRONIC INSTRUMENT HANDBOOK
Di Giacomo • VLSI HANDBOOK
Fink and Beaty • STANDARD HANDBOOK FOR ELECTRICAL ENGINEERS
Fink and Christiansen • ELECTRONICS ENGINEERS' HANDBOOK
Flanagan • HANDBOOK OF TRANSFORMER APPLICATIONS
Harper • HANDBOOK OF ELECTRONIC SYSTEMS DESIGN
Harper • HANDBOOK OF THICK FILM HYBRID MICROELECTRONICS
Inglis • ELECTRONIC COMMUNICATIONS HANDBOOK
Johnson and Jasik • ANTENNA ENGINEERING HANDBOOK
Juran • QUALITY CONTROL HANDBOOK
Kaufman and Seidman • HANDBOOK OF ELECTRONICS CALCULATIONS
Kaufman and Seidman • HANDBOOK FOR ELECTRONICS ENGINEERING TECHNICIANS
Stout • HANDBOOK OF MICROPROCESSOR DESIGN AND APPLICATIONS
Stout and Kaufman • HANDBOOK OF MICROCIRCUIT DESIGN AND APPLICATION
Stout and Kaufman • HANDBOOK OF OPERATIONAL AMPLIFIER CIRCUIT DESIGN
Williams • DESIGNER'S HANDBOOK OF INTEGRATED CIRCUITS
Williams and Taylor • ELECTRONIC FILTER DESIGN HANDBOOK

Encyclopedias

CONCISE ENCYCLOPEDIA OF SCIENCE AND TECHNOLOGY
ENCYCLOPEDIA OF ELECTRONICS AND COMPUTERS
ENCYCLOPEDIA OF ENGINEERING

Dictionaries

DICTIONARY OF SCIENTIFIC AND TECHNICAL TERMS
DICTIONARY OF ELECTRICAL AND ELECTRONIC ENGINEERING
DICTIONARY OF ENGINEERING

Other

Antognetti • POWER INTEGRATED CIRCUITS
Chryssis • HIGH-FREQUENCY SWITCHING POWER SUPPLIES
Grossner • TRANSFORMERS IN ELECTRONIC CIRCUITS
Mitchell • DC/DC SWITCHING REGULATOR ANALYSIS
Rombaut • POWER ELECTRONIC CONVERTERS: AC/AC CONVERSION
Seguier • POWER ELECTRONIC CONVERTERS: AC/DC CONVERSION

*For more information about other McGraw-Hill materials,
call 1-800-2-MCGRAW in the United States. In other
countries, call your nearest McGraw-Hill office.*

GEORGE CHRYSSIS

HIGH-FREQUENCY SWITCHING POWER SUPPLIES: THEORY AND DESIGN

SECOND EDITION

McGRAW-HILL PUBLISHING COMPANY   NEW YORK   ST. LOUIS   SAN FRANCISCO
AUCKLAND BOGOTÁ HAMBURG LONDON MADRID MEXICO MILAN MONTREAL NEW DELHI
PANAMA   PARIS   SÃO PAULO   SINGAPORE   SYDNEY   TOKYO   TORONTO

Library of Congress Cataloging-in-Publication Data

Chryssis, George.

  High-frequency switching power supplies.

  Includes index.
  1. Electronic apparatus and appliances—Power supply.
2. Microelectronics—Power supply. I. Title.
TK7868.P6C47 1989   621.381′044   88-13681
ISBN 0-07-010951-6

1234567890   DOC/DOC   89432109

ISBN 0-07-010951-6

*The editors for this book were Daniel A. Gonneau and Marci Nugent, the
designer was the Riverside Graphic Studio, Inc., and the production
supervisor was Suzanne W. Babeuf. It was set in Caledonia by Techna
Type, Inc.*

Printed and bound by R. R. Donnelley & Sons Company.

*For more information about other McGraw-Hill materials,
call 1-800-2-MCGRAW in the United States. In other
countries, call your nearest McGraw-Hill office.*

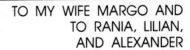

TO MY WIFE MARGO AND
TO RANIA, LILIAN,
AND ALEXANDER

CONTENTS

FOUR

THE POWER TRANSISTOR IN CONVERTER DESIGN

SEVEN
SWITCHING REGULATOR CONTROL CIRCUITS 183

EIGHT
SWITCHING POWER SUPPLY ANCILLARY, SUPERVISORY, AND PERIPHERAL CIRCUITS AND COMPONENTS 209

PREFACE

The rapid advancement of microelectronics in recent years has demanded the development of sophisticated, efficient, lightweight power systems which have a high power-to-volume density ratio with no compromise in performance. The high-frequency switching power supply meets these demands, and recently it has become the prime powering source in the majority of modern electronic designs.

Because the complexity of the power supply is increasing, dedicated engineering departments and highly skilled engineers are required to undertake their design and development. Unfortunately, very few engineers are college-trained to become power supply engineers, and the ones who get involved and make it a career do so either due to circumstantial involvement or by demand.

On the other hand, very few colleges and universities offer any power electronic courses with the emphasis on the design of switching power supplies and magnetics. Consequently, college students or practicing engineers who want to pursue a career in this truly exciting and fascinating field must enrich their knowledge by studying material mostly in the form of application notes, published by different electronic companies in promoting their products, or technology articles published in professional magazines. The need for a comprehensive, coherent, and easy to understand text, which blends both theory and practice and also covers most of the latest developments in the switching power supply field, promoted the writing of this book.

This book is intended to be used by either the engineering student or the practicing engineer who wants step-by-step instruction to the theory and design of switching power supplies. It compiles the knowledge of all those who have worked in this field for years in order to help those just starting. It includes enough theory to make the reader aware of the results, but long mathematical derivations are limited. The end result of the theory and its application in the design aspect is emphasized.

This Second Edition of the book follows the format set forth in the First Edition. However, certain chapters have been enhanced and expanded with new material. The net result is a more comprehensive and updated presentation of high-frequency switching power supply design, which takes the reader from the 20-kHz region to today's megahertz region.

The first chapter of the book describes the building blocks of a complete switcher, and each of the subsequent chapters describes and analyzes each building block in detail. The basic design equations are given, but long mathematical derivations are omitted. Numerous examples are presented that enhance and reinforce the theory with practical designs.

The book describes all the basic classical switching power supply topologies, and new topologies are also presented. The switching power transistor, both bipolar and MOSFET, is extensively analyzed. Subjects such as snubber circuit designs and base drive designs are extensively covered. Other semiconductor devices, such as the gate turn-off (GTO), synchronous rectifiers, fast recovery rectifiers, and Schottky rectifiers are also presented in depth.

The analysis and design of magnetic components, such as high-frequency power transformers, power inductors, and magnetic amplifiers along with application design examples are covered.

Power control semiconductor integrated circuits (ICs) are covered, and a number of representative popular commercial ICs are descsribed as a reference.

Chapter 9 is devoted to the important subject of power supply feedback loop stability. The analysis and design of feedback amplifiers is presented in a practical, easy to understand way, simplifying this traditionally difficult to understand subject.

Chapter 10 discusses the importance of electromagnetic and radio frequency interference (EMI-RFI) and presents circuits for its suppression.

Chapter 11 covers national and international safety requirements, and it presents and explains safety design standards such as UL, CSA, VDE, and IEC.

The First Edition of the book was received with enthusiasm, and I was gratified to receive both compliments and constructive suggestions. These suggestions prompted the writing of the Second Edition, which includes material asked for by readers.

It is my hope that this edition will become a valuable reference to all those seeking knowledge and understanding of the ever-expanding field of high-frequency switching power supply design.

I would again like to thank my family for their support and understanding during the writing of both editions of the book, as well as all those who have contributed to the enhancement of our profession and the power supply field in general.

Also I would like to thank Ellen Dalmus and Claudia Mungle for typing the manuscript for the First Edition and Barbara Stone and Jami Schmid for their contribution in typing the manuscript for the Second Edition.

George C. Chryssis

ONE

THE SWITCHING POWER SUPPLY: AN OVERVIEW

1-0 INTRODUCTION

As the integrated semiconductor technology becomes more advanced, system designers as well as electronic manufacturers are emphasizing size and weight as important features of their products.

Traditionally, the bulkiest part of a system is the power supply, with its heavy isolation transformer, heat sinks, and cooling fans, as in the case of series-pass linear designs.

The trend therefore in recent years has been toward the development of high-efficiency, lightweight, and compact power supplies to complement the new system designs. The high-frequency switching power supply is the obvious solution.

However, this new type of power supply is much more sophisticated than its linear counterpart. It requires knowledge in analog electronic design and magnetic component design, along with logic and control design.

The task of power supply design is not a side project anymore. The switching power supply has spawned new exciting interest in the power electronics field, and the term "power supply engineer" has been redefined and given new respect. New research has been undertaken by the industry and academia, to push the frontiers of this truly fascinating field.

The advancements in the field have been rapid and rewarding. The power supply is finally following the advancement of the rest of the electronics. Power supplies are becoming smaller, more efficient, extremely compact, and cost-effective. The trend is for higher switching frequencies, above the traditional 20-kHz and up to the megahertz region. Commercial power supplies working at 1 MHz are already available from a number of manufacturers, with more to come.

Truly this is only the beginning.

1-1 THE LINEAR POWER SUPPLY

The linear power supply is already a mature technology, which has been used since the dawn of electronics. Whether this type of power supply incorporates tubes or semiconductors, its construction and operation are essentially the same.

Figure 1-1 shows the simplified block diagram of a series-pass linear regulated power supply. In this type of power supply, a low-frequency, 50- or 60-Hz transformer is used to step down the ac mains to a lower voltage of the same frequency. This secondary voltage is in turn rectified and filtered, and the resulting dc is fed into a series-pass active element.

By sampling a portion of the output voltage and comparing it to a fixed reference voltage, the series-pass element is used as a form of "variable resistor" to control and regulate the output voltage. However, this mode of operation dissipates a large amount of power in the form of heat, consequently lowering the efficiency of the power supply to 40 or 50 percent.

Although the linear power supply in general has a tight regulation band along with very low output noise and ripple, the disadvantages are obvious.

As we mentioned, because of its low efficiency, usually bulky and expensive heat sinks and cooling fans are needed, and large isolation power transformers are used to step down the ac input voltage. Hence, this type of power supply tends to be bulky, heavy, and almost unfit for today's compact electronic systems.

Other disadvantages of the linear power supply are its relatively narrow input voltage range, normally ±10 percent, and its very low output hold-up time, about 1 ms.

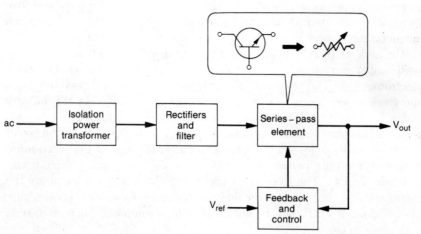

FIGURE 1-1 Block diagram of a series-pass regulated linear power supply.

1-2 THE OFF-THE-LINE SWITCHING REGULATED POWER SUPPLY

The disadvantages of the linear power supply are greatly reduced or eliminated by the regulated switching power supply.

Figure 1-2 shows a simplified block diagram of a high-frequency off-the-line switching power supply. In this scheme the ac line is directly rectified and filtered to produce a raw high-voltage dc, which in turn is fed into a switching element. The switch is operating at the high frequencies of 20 kHz to 1 MHz, chopping the dc voltage into a high-frequency square wave. This square wave is fed into the power isolation transformer, stepped down to a predetermined value and then rectified and filtered to produce the required dc output.

A portion of this output is monitored and compared against a fixed reference voltage, and the error signal is used to control the on-off times of the switch, thus regulating the output.

Since the switch is either on or off, it is dissipating very little energy, resulting in a very high overall power supply efficiency of about 70 to 80 percent. Another advantage is the power transformer size which can be quite small due to high operating frequency. Hence the combination of high efficiency (i.e., no large heat sinks) and relatively small magnetics, results in compact, lightweight power supplies, with power densities up to 30 W/in.[3] versus 0.3 W/in.[3] for linears. Coupled with very wide input voltage range, 90 to 260 V ac, and very good hold-up time, typically 25 ms, the switcher has become the choice for electronic system designers.

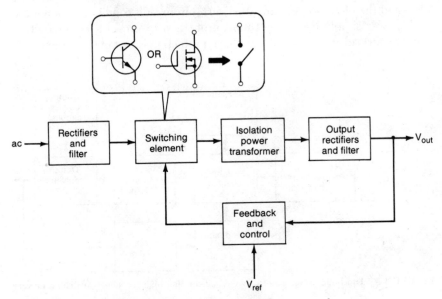

FIGURE 1-2 The basic off-the-line switching regulated power supply.

Of course, there are certain disadvantages associated with the switching power supply, such as higher output noise and ripple, EMI/RFI generation, and higher design complexity. However, with careful design these problems may be greatly reduced or eliminated.

1-2-1 The Complete Off-the-Line Power Supply Building Blocks

The building blocks of a typical high-frequency off-the-line switching power supply are depicted in Fig. 1-3.

The purpose of this book is to analyze each building block in a logical sequence, providing the reader with the understanding and with the tools needed to undertake the design of switching power supplies. As the name implies, the input rectification in a switching power supply is done directly off-the-line, without the use of the low-frequency isolation power transformer, as in the case of the linear power supply.

The basic operation of the switching power supply was described in the previous section. However, the complete block diagram shown in Fig. 1-3 contains other important sections such as the ac line RFI filter block, the ancillary and supervisory blocks, and the I/O isolation block.

The EMI/RFI filter could be either part of the power supply or external to it, and it is generally designed to comply to national or international specifications, such as the FCC class A or class B and VDE-0871.

The ancillary and supervisory circuits are used to protect the power supply—and the electronic circuits it powers—from fault conditions. Generally each power supply has current-limit protection to prevent its destruction during overload conditions. Overvoltage protection is part of the supervisory

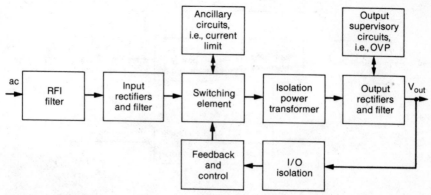

FIGURE 1-3 The building blocks of a typical off-the-line high-frequency switching power supply.

circuits used to protect the load from power supply failures. It is important to note that although in a linear power supply, overvoltage conditions were common during failure of the series-pass element (i.e., shorted), in a switching power supply, failure of the switching element normally results in a nonoutput condition. However, the output of the switcher will go high if the feedback loop is opened.

Input/output isolation is essential to an off-the-line switcher. The isolation may be optical or magnetic, and it should be designed to comply to UL/CSA or VDE/IEC safety standards. Thus the UL and CSA require 1000 V ac isolation voltage withstand while VDE and IEC require 3750 V ac. Consequently the power transformer has to be designed to the same safety isolation requirements also.

REFERENCES

1. Bird, B. M., and K. G. King: "An Introduction to Power Electronics," Wiley, New York, 1983.

2. Chryssis, G., and B. Friedman: Glossary of Power Supply Terms, *EPM*, July 1979.

3. Hnatek, E. R.: "Design of Solid State Power Supplies," 2d ed., Van Nostrand Reinhold, New York, 1981.

4. Middlebrook, R. D., and S. Ćuk: "Advances in Switched Mode Power Converters," vols. 1 and 2, Teslaco, Pasadena, Calif., 1981.

5. Mitchell, D. M.: "DC/DC Switching Regulator Analysis," McGraw-Hill, New York, 1988.

6. Pressman, A. I.: "Switching and Linear Power Supply, Power Converter Design," Hayden, Rochelle Park, N.J., 1977.

7. Severns, R. P., and G. E. Bloom: "Modern DC-to-DC Switchmode Power Converter Circuits," Van Nostrand Reinhold, New York, 1985.

TWO

THE INPUT SECTION

2-0 THE VOLTAGE DOUBLER TECHNIQUE

As we mentioned previously, an off-the-line switching power supply rectifies the ac line directly without requiring a low-frequency, line isolation transformer between the ac mains and the rectifiers. Since in most of today's electronic equipment the manufacturer is generally addressing an international market, the power supply designer must use an input circuit capable of accepting all world voltages, normally 90 to 130 V ac or 180 to 260 V ac.

Figure 2-1 shows a realization of such a circuit by using the voltage doubler technique. When the switch is closed, the circuit may be operated at a nominal line of 115 V ac. During the positive half cycle of the ac, capacitor C_1 is charged to the peak voltage, 115 V ac $\times$ 1.4 = 160 V dc, through diode D_1. During the negative half cycle, capacitor C_2 is charged to 160 V dc through diode D_4. Thus, the resulting dc output will be the sum of the voltages across $C_1 + C_2$, or 320 V dc. When the switch is open, D_1 to D_4 form a full-bridge rectifier capable of rectifying a nominal 230-V ac line and producing the same 320-V dc output voltage.

2-1 COMPONENT SELECTION AND DESIGN CRITERIA

2-1-1 Input Rectifiers

When choosing either a bridge rectifier assembly or discrete rectifiers, the designer must look up the following important specifications:

1. Maximum forward rectification current capability. This figure depends primarily on the power level of the switching power supply design, and the selected diode must have at least twice the steady-state current capacity of the calculated value.

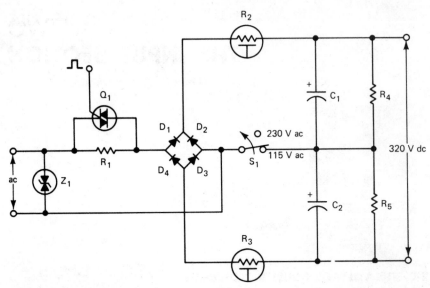

FIGURE 2-1 This circuit may be used for either 115- or 230-V ac input, depending on the position of the switch. Inrush current limiting, input transient protection, and discharge resistors are also shown.

2. Peak inverse voltage (PIV) blocking capability. Since these rectifiers are used in high-voltage environments, they must have a high PIV rating, normally 600 V or more.

3. High surge current capabilities to withstand the peak currents associated with turn-on.

2-1-2 Input Filter Capacitors

Proper calculation and selection of the input rectifier filter capacitors is very important, since this will influence the following performance parameters: the low-frequency ac ripple at the output of the power supply and the holdover time. Normally high-grade electrolytic capacitors with high ripple current capacity and low ESR are used with a working voltage of 200 V dc minimum. Resistors R_4 and R_5, shown in Fig. 2-1 shunting the capacitors, provide a discharge path when the supply is switched off.

The formula to calculate the filter capacitor is given by

$$C = \frac{It}{\Delta V} \qquad (2\text{-}1)$$

where C = capacitance, μF
$\qquad I$ = load current, A
$\qquad t$ = time the capacitor must supply currents, ms
$\qquad \Delta V$ = allowable peak-to-peak ripple, V

EXAMPLE 2-1

Calculate the value of the input filter capacitors of a 50-W switching power supply working at 115 V ac, 60 Hz.

SOLUTION

The first step is to calculate the dc load current. Assume a worst-case efficiency of 70 percent for the power supply. Then for a 50-W output the input power is

$$P_{in} = \frac{P_{out}}{\eta} = \frac{50}{0.7} = 71.5 \text{ W}$$

Using the voltage doubler technique (see Fig. 2-1), at 115 V ac, the dc output will be $2(115 \times 1.4) = 320$ V dc. Therefore, the load current will be $I = P/E = 71.5/320 = 0.22$ A. Now assume the design can tolerate a ripple of 30 V peak-to-peak and that the capacitor has to maintain the voltage level for every half cycle, i.e., one-half the line frequency or for about 8 ms for a 60-Hz ac line frequency. Then using Eq. 2-1,

$$C = \frac{0.22(8 \times 10^{-3})}{30} = \frac{1.76 \times 10^{-3}}{30} = 58 \times 10^{-6} \text{ F} = 58 \text{ } \mu\text{F}$$

We choose a standard value of 50 μF.

Since in the voltage doubler configuration $C = C_1 + C_2$, then $C_1 = C_2 = 100 \text{ } \mu$F, which is the capacity needed for this 50-W design.

2-2   INPUT PROTECTIVE DEVICES

2-2-1   Inrush Current

An off-the-line switching power supply may develop extremely high peak inrush currents during turn-on, unless the designer incorporates some form of current limiting in the input section. These currents are caused by the charging of the filter capacitors, which at turn-on present a very low impedance to the ac lines, generally only their ESR. If no protection is employed, these surge currents may approach hundreds of amperes.

Two methods are widely employed in introducing an impedance to the ac line at turn-on and in limiting the inrush current to a safe value. One is using a resistor-triac arrangement, and the other using negative temperature coefficient (NTC) thermistors. Figure 2-1 shows how these elements may be employed in a power supply.

The Resistor-Triac Technique   Using this inrush current limiting technique, a resistor is placed in series with the ac line. The resistor is shunted by a triac which shorts the resistor out when the input filter capacitors have been fully charged. This arrangement requires a trigger circuit which will fire the triac on when some predetermined conditions have been met. Care must be taken in choosing and heat-sinking the triac so that it can handle the full input current when it is turned on.

The Thermistor Technique   This method uses NTC thermistors placed on either the ac lines or the dc buses after the bridge rectifiers, as shown in Fig. 2-1.

The resistance-temperature characteristics and the relationship of the temperature coefficient α of the NTC thermistor are shown in Fig. 2-2. When the power supply is switched on, the resistance of the thermistor(s)

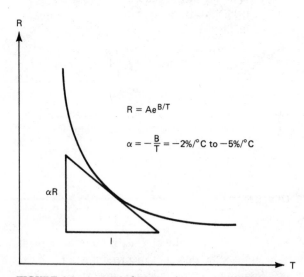

$$R = Ae^{B/T}$$

$$\alpha = -\frac{B}{T} = -2\%/°C \text{ to } -5\%/°C$$

FIGURE 2-2 An NTC thermistor's resistance drops drastically as the temperature increases; α is the temperature coefficient of the thermistor, expressed in percentage per degree centigrade.

is essentially the only impedance across the ac line, thus limiting the inrush current.

As the capacitors begin to charge, current starts to flow through the thermistors, heating them up. Because of their negative temperature coefficient, as the thermistors heat up their resistance drops. When the thermistors are properly chosen, their resistance at steady-state load current will be a minimum, thus not affecting the overall efficiency of the power supply.

2-2-2 Input Transient Voltage Protection

Although the ac mains are nominally rated at 115 V ac or 230 V ac, it is common for high-voltage spikes to be induced, caused by nearby inductive switching or natural causes such as electrical storms or lightning. During severe thunderstorm activity, voltage spikes in the order of 5 kV are not uncommon.

On the other hand, inductive switching voltage spikes may have an energy content

$$W = \tfrac{1}{2}LI^2 \qquad\qquad (2\text{-}2)$$

where L is the leakage inductance of the inductor, and I is the current flowing through the winding.

Therefore, although these voltage spikes may be short in duration, they may carry enough energy to prove fatal for the input rectifiers and the switching transistors, unless they are successfully suppressed.

The most common suppression device used in this situation is the metal oxide varistor (MOV) transient voltage suppressor, and it may be used as shown in Fig. 2-1 across the ac line input. This device acts as a variable impedance; that is, when a voltage transient appears across the varistor, its impedance sharply decreases to a low value, clamping the input voltage to a safe level. The energy in the transient is dissipated in the varistor. The following procedure gives a guide in selecting these devices:

1. Choose the ac voltage rating of the MOV to be about 10 to 20 percent greater than the maximum steady-state circuit value.

2. Calculate or estimate the maximum transient energy in joules that the circuit may experience.

3. Make sure that the maximum peak surge current of the device will be properly rated.

After all of the above have been established, the proper MOV may be selected from the manufacturer's data sheets.

REFERENCES

1. Chryssis, G.: Power Supplies: A Switching Alternative, *Design News*, May 21, 1979.

2. Hnatek, E. R.: "Design of Solid State Power Supplies," 2d ed., Van Nostrand Reinhold, New York, 1981.

3. Pressman, A. I.: "Switching and Linear Power Supply, Power Converter Design," Hayden, Rochelle Park, NJ, 1977.

THREE

TYPES OF POWER CONVERTERS

3-0 DEFINITIONS AND DIMENSIONING

Although there are numerous converter circuits described by a number of authors and researchers, basically all of them are related to three classical circuits known as the "flyback or buck-boost," the "forward or buck," and the "push-pull or buck-derived" converter. A model of the flyback converter is illustrated in Fig. 3-1, and operation of the circuit is as follows.

When the switch S is closed (Fig. 3-1a), current flows through inductor L, storing energy. Because of the voltage polarity, diode D is reverse-biased, thus no voltage is present across the load R_L. When the switch S is open (Fig. 3-1b), inductor L reverses polarity because of the collapsing magnetic field, forward-biasing diode D, and inducing a current flow I_C in the polarity shown. Thus, an output voltage of opposing polarity to the input voltage appears across R_L. Since the switch commutates continuous inductor current alternately between input and output, both currents are pulsating in form. Therefore, in the buck-boost converter, energy is stored in the inductor during the switch on period; then this energy is transferred to the load during the flyback or the switch off period.

Figure 3-2 illustrates the operation of a forward converter. When the switch S is closed, current I flows in a forward manner through inductor L, producing an output voltage across the load with the polarity shown in Fig. 3-2a. Diode D is also reverse-biased due to the direction of the input voltage polarity. When the switch S opens (Fig. 3-2b), the magnetic field in L changes polarity, forward-biasing diode D and producing a current through capacitor C, as shown. Therefore, the output voltage polarity across R_L remains the same. Diode D is often called a "free-wheeling" or "flywheel" diode.

Because of this switching action, the output current is continuous and therefore nonpulsating. In contrast the input current is discontinuous, hence

13

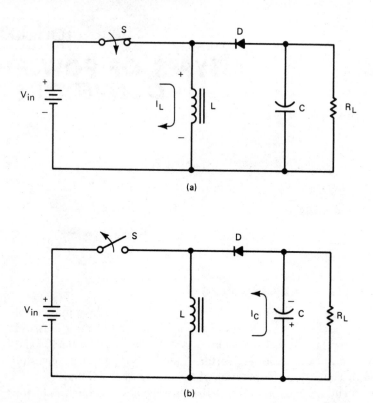

FIGURE 3-1 Model of the flyback or buck-boost converter. (*a*) Switch closed, (*b*) switch open.

pulsating, flowing when the switch is closed and abruptly interrupted when the switch is open.

Last, Fig. 3-3 shows the push-pull converter, which in reality consists of the two forward converters operating in a push-pull (or more correctly push-push) action, with alternate closing and opening of either switch S_1 or S_2. This circuit is also known as buck-derived.

3-1 THE ISOLATED FLYBACK CONVERTER

The model of the flyback converter shown in Fig. 3-1 has no safety isolation between input and output. An off-the-line switching power supply normally requires mains isolation in the form of a transformer. To be more precise, although in the diagram the isolation element appears in the form of a transformer, its action is that of a choke and therefore it is more correctly referred to as transformer-choke.

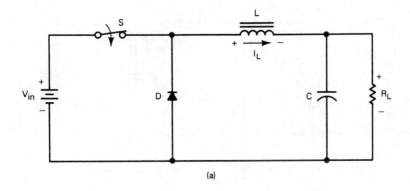

(a)

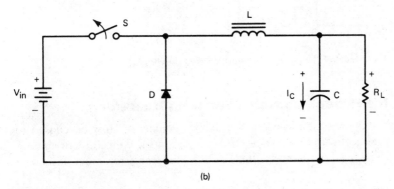

(b)

FIGURE 3-2 Model of the forward or buck converter. (*a*) Switch closed, (*b*) switch open.

Figure 3-4 depicts an isolated flyback converter with the associated steady-state waveforms. The circuit operates as follows. When transistor Q_1 is on, primary current starts to build up in the primary winding, storing energy. Due to the opposite polarity arrangement between the input and output windings of the transformer-choke, there is no energy transferred to the load since diode D is reverse-biased.

When the transistor is turned off, the polarity of the windings reverses due to the collapsing magnetic field. Now diode D is conducting, charging the output capacitor C and delivering current I_L to the load.

Since the isolation element acts as both a transformer and a choke, no additional inductor is needed on the output section of the flyback converter. In practice, though, a small inductor may be necessary between the rectifier and the output capacitor in order to suppress high-frequency switching noise spikes.

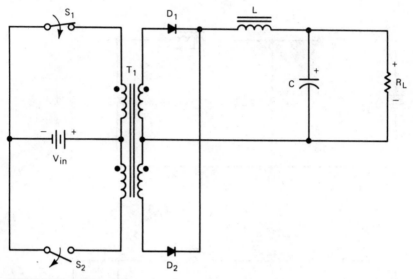

FIGURE 3-3 Model of the push-pull or buck-derived converter.

3-1-1 The Flyback Converter Switching Transistor

The switching transistor used in the flyback converter must be chosen to handle peak collector voltage at turn-off and peak collector currents at turn-on. The peak collector voltage which the transistor must sustain at turn-off is

$$V_{CE,\max} = \frac{V_{in}}{1 - \delta_{\max}} \tag{3-1}$$

where V_{in} is the dc input voltage, and $\delta_{\max}$ is the maximum duty cycle.

Equation 3-1 tells us by examination that in order to limit the collector voltage to a safe value, the duty cycle must be kept relatively low, normally below 50 percent, i.e., $\delta_{\max} < 0.5$. In practice $\delta_{\max}$ is taken at about 0.4, which limits the peak collector voltage $V_{CE,\max} \leq 2.2V_{in}$, and, therefore, transistors with working voltages above 800 V are usually used in off-the-line flyback converter designs.

The second design criterion which the transistor must meet is the working collector current at turn-on, given by

$$I_C = \frac{I_L}{n} = I_P \tag{3-2}$$

where I_P is the primary transformer-choke peak current, n is the primary-to-secondary turns ratio, and I_L is the output load current.

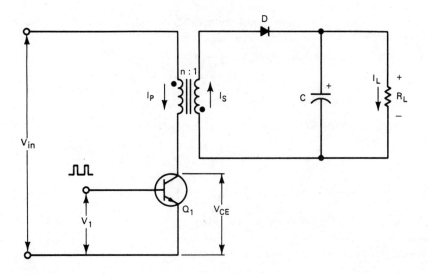

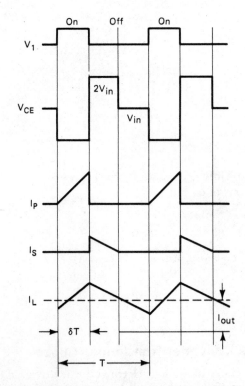

FIGURE 3-4 The isolated flyback converter and its associated waveforms.

To derive an expression of the peak working collector current in terms of converter output power and input voltage, the following equation may be written for the energy transferred in the choke:

$$P_{out} = \left(\frac{LI_P^2}{2T}\right)\eta \tag{3-3}$$

where η (eta) is the efficiency of the converter.

The voltage across the transformer may be expressed as

$$V_{in} = \frac{L\,di}{dt} \tag{3-4}$$

If we assume $di = I_P$ and $1/dt = f/\delta_{max}$, then Eq. 3-4 may be rewritten as

$$V_{in} = \frac{LI_P f}{\delta_{max}} \tag{3-5}$$

or

$$L = \frac{V_{in}\delta_{max}}{I_P f} \tag{3-6}$$

Substituting Eq. 3-6 into Eq. 3-3 we get

$$P_{out} = \left(\frac{V_{in}f\delta_{max}I_P^2}{2fI_L}\right)\eta = \tfrac{1}{2}\eta V_{in}\delta_{max}I_P$$

Solving for I_P,

$$I_P = \frac{2P_{out}}{\eta V_{in}\delta_{max}} \tag{3-7}$$

Now substituting Eq. 3-7 into Eq. 3-2 we get the expression for the transistor working current in terms of output power

$$I_C = \frac{2P_{out}}{\eta V_{in}\delta_{max}} \tag{3-8}$$

To simplify Eq. 3-8 even more, assuming a converter efficiency of 0.8 (80 percent) and a duty cycle $\delta_{max} = 0.4$ (40 percent), then

$$I_C = \frac{6.2P_{out}}{V_{in}} \tag{3-9}$$

3-1-2 The Flyback Converter Transformer-Choke

Since the transformer-choke of the flyback converter is driven in one direction only of the B-H characteristic curve, it has to be designed so that it will not saturate. An extensive analysis and design is given in Chap. 5. Here it suffices to say that a core with a relatively large volume and air gap must be used.

The effective transformer-choke volume is given by

$$\text{Volume} = \frac{\mu_0 \mu_e I_{L,\max}^2 L_{\text{out}}}{B_{\max}^2} \qquad (3\text{-}10)$$

where $I_{L,\max}$ = determined by load current
μ_e = relative permeability of the chosen core material
$B_{\max}$ = maximum flux density of the core

The relative permeability μ_e must be chosen to be large enough to avoid excessive temperature rise in the core due to restricting core and wire size and therefore copper and core losses.

3-1-3 Variations of the Basic Flyback Converter

As was discussed with the basic flyback circuit, the collector voltage of the switching transistor must sustain at least twice the input voltage at turn-off. In cases where the voltage value is too high to use a commercial transistor type, the two-transistor flyback converter may be used, as shown in Fig. 3-5. This circuit uses two transistors, which are switched on or off simultaneously. Diodes D_1 and D_2 act as clamping diodes restricting the maximum collector voltage of the transistors to V_{in}. Thus lower voltage transistors may be used to realize this design, but at the expense of three extra components, i.e., Q_2, D_1, and D_2.

An advantage of the flyback circuit is the simplicity by which a multiple output switching power supply may be realized. This is because the isolation element acts as a common choke to all outputs, thus only a diode and a capacitor are needed for an extra output voltage. Figure 3-6 illustrates a practical circuit.

3-2 THE ISOLATED FORWARD CONVERTER

At first glance, the isolated forward converter circuit resembles that of the flyback converter, but there are some definite and distinct differences between the two circuit topologies and their operation. Figure 3-7 shows the basic forward converter and its associated waveforms.

Since the isolation element in the forward converter is a pure transformer, a second inductive energy storage element L is required at the output for proper and efficient energy transfer. Notice also that the primary and secondary windings of the transformer have the same polarity, i.e., the dots

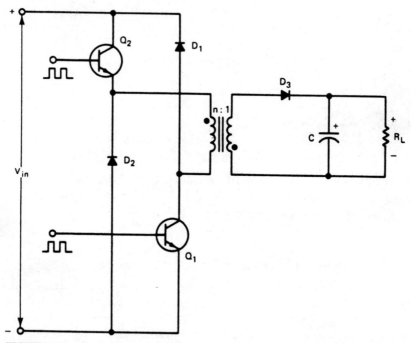

FIGURE 3-5 The two-transistor flyback circuit limits the collector voltage of each transistor to V_{in}.

are at the same winding ends. Function of the circuit is as follows. When Q_1 turns on, current builds up in the primary winding, storing energy. Because of the same polarity of the transformer secondary winding, this energy is forward-transferred to the output and also stored in inductor L through diode D_2 which is forward-biased. Diode D_3 is back-biased. When Q_1 turns off, the transformer winding voltage reverses, back-biasing diode D_2. Now, the flywheel diode D_3 is forward-biased, conducting current in the output loop and delivering energy to the load through inductor L.

The tertiary winding and diode D_1 provide transformer demagnetization when Q_1 is switched off by returning the transformer magnetic energy into the output dc bus. The dark areas on the waveforms of Fig. 3-7 show the magnetizing-demagnetizing current, given as

$$I_{mag} = \frac{T\delta_{max}V_{in}}{L} \qquad (3\text{-}11)$$

where $T\delta_{max}$ is the period when transistor Q_1 is on, and L is the output inductance in microhenrys.

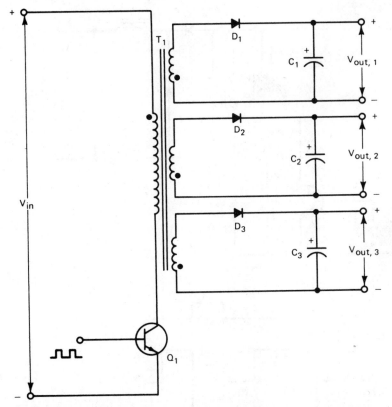

FIGURE 3-6 Multiple outputs may be easily derived using a flyback converter. Both positive and negative voltages are possible using extra output windings, a diode, and a smoothing capacitor.

3-2-1 The Forward Converter Switching Transistor

Because of the tertiary winding and diode D_1 (Fig. 3-7), the voltage across transistor Q_1 at turn-off is limited to

$$V_{CE,\max} = 2V_{in} \qquad (3\text{-}12)$$

The waveforms also show that the peak collector voltage of $2V_{in}$ is maintained for as long as D_1 conducts, that is, for a period of $T\delta_{\max}$. Also by inspection of the same waveforms we can see that the transistor collector current at turn-on will have a value equal to that derived for the flyback converter plus the net amount of the magnetization current. Therefore, peak collector current in the transistor may be written as

$$I_C = \frac{I_L}{n} + \frac{T\delta_{\max}V_{in}}{L} \qquad (3\text{-}13)$$

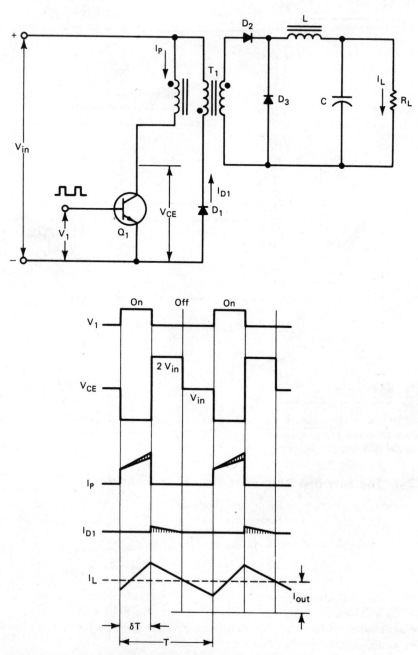

FIGURE 3-7 Realization of the isolated forward converter and its associated waveforms. The cross-hatched areas depict transformer magnetizing currents.

where n = primary-to-secondary turns ratio
I_L = output inductor current, A
$T\delta_{max}$ = period transistor is on
L = output inductance, μH

From the expression

$$V_{out} = \frac{\delta_{max}V_{in}}{n} \tag{3-14}$$

or

$$V_{in} = \frac{nV_{out}}{\delta_{max}} \tag{3-15}$$

Equation 3-13 may be rewritten as

$$I_C = \frac{I_L}{n} + \frac{nTV_{out}}{L} \tag{3-16}$$

Assuming that the magnetizing current term nTV_{out}/L is very small compared to the peak collector current, for all practical purposes I_C may be derived as shown in Sec. 3-1-1 to be

$$I_C = \frac{I_L}{n} \approx \frac{6.2P_{out}}{V_{in}} \tag{3-17}$$

3-2-2 The Forward Converter Transformer

Care must be taken when designing the transformer of the forward converter to choose the proper core volume and air gap to avoid saturation. The transformer equations given in Chap. 5 may be used to design the forward transformer. The core volume of the transformer is given by

$$\text{Volume} = \frac{\mu_0\mu_e I_{mag}^2 L}{B_{max}^2} \tag{3-18}$$

where

$$I_{mag} = \frac{nTV_{out}}{L} \tag{3-19}$$

It should also be noted that the duty cycle of the switch δ_{max} must be kept below 50 percent, so that when the transformer voltage is clamped through the tertiary winding, the integral of the volt-seconds between the input voltage, when Q_1 is on, and the clamping level, when Q_1 is off, amounts to zero. Duty cycles above 50 percent, i.e., $\delta > 0.5$, will upset the volt-seconds balance, driving the transformer into saturation, which in turn produces high collector current spikes that may destroy the switching transistor.

Although the clamping action of the tertiary winding and the diode limit the transistor peak collector voltage to twice the dc input, care must be taken during construction to couple the tertiary winding tightly to the primary (bifilar wound) to eliminate fatal voltage spikes caused by leakage inductance.

3-2-3 Variations of the Basic Forward Converter

In cases where the input voltage is too high, the two-transistor forward converter may be used, as in the flyback case. The two transistors are switched on or off simultaneously, but the transistor voltage does not rise above V_{in}. Figure 3-8 shows this circuit variation.

The forward converter may also be used to derive multiple output voltages, although this circuit requires an extra diode and choke in the output section of each derived voltage. It should be noted here that the flywheel diode must have at least the same current rating as the main rectifying diode,

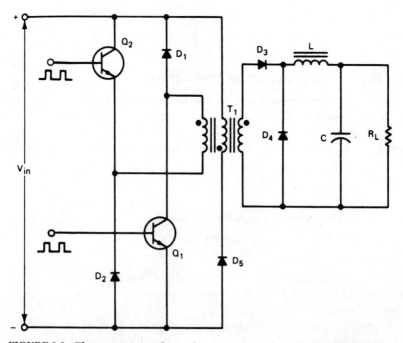

FIGURE 3-8 The two-transistor forward converter limits the collector voltage of each transistor to V_{in} because of the clamping action of D_1 and D_2.

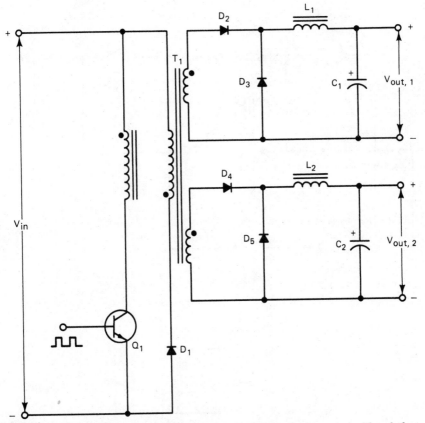

FIGURE 3-9 The forward transformer suits itself well to multiple outputs, although the circuit realization is more expensive than that of the flyback converter because of the extra diode and choke in each output.

since it provides full output current when the transistor is off. Figure 3-9 shows a practical multiple-output forward converter.

3-3 THE PUSH-PULL CONVERTER

The push-pull converter is really an arrangement of two forward converters working in antiphase. Since both halves of the push-pull converter are delivering power to the load at each half cycle, a more appropriate name might have been the push-push converter, but over the years, the term push-pull has prevailed.

Figure 3-10 shows the basic conventional push-pull configuration and its associated waveforms. From the waveforms we can see that because of the

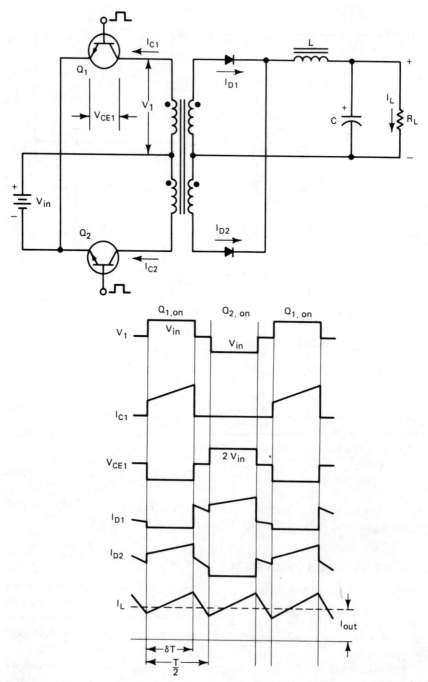

FIGURE 3-10 The push-pull converter and its associated waveforms.

presence of the two sets of switching transistors and output diodes, the average current in each set is reduced to 50 percent over the equivalent forward converter. Note that in the interval between transistor conduction, diodes D_1 and D_2 conduct simultaneously, essentially shorting the secondary isolation transformers and delivering power to the output, acting as flywheel diodes.

The output voltage of this converter may be derived as

$$V_{out} = \frac{2\delta_{max}V_{in}}{n} \qquad (3\text{-}20)$$

The value of δ_{max} in Eq. 3-20 must remain below 0.5 in order to avoid simultaneous conduction of the switching transistors and therefore destruction. Assuming $\delta_{max} = 0.4$, then Eq. 3-20 may be rewritten as

$$V_{out} = \frac{0.8V_{in}}{n} \qquad (3\text{-}21)$$

where n is the primary-to-secondary turns ratio.

3-3-1 The Push-Pull Converter Transformer

As we noted when discussing flyback and forward converters, their transformers exploit only one-half of the $B\text{-}H$ characteristic curve and therefore are bulky and have an air gap. Assuming in the push-pull converter that the conduction times of the two transistors are equal (or they are forced to be equal) the transformer will use both halves of the $B\text{-}H$ curve and the volume of the core will be halved. An air gap may not be necessary.

The volume of the transformer is given by

$$\text{Volume} = \frac{4\mu_0\mu_e I_{mag,L}^2}{B_{max}^2} \qquad (3\text{-}22)$$

where $I_{mag} = nV_{out}T/4L$ is the magnetizing current.

In Chap. 3 an extensive design analysis is given for the family of push-pull–based converters.

3-3-2 The Push-Pull Converter Transistors

Since each half of the push-pull converter in essence is a forward converter, the voltage across each transistor at turn-off is limited to

$$V_{CE,max} = 2V_{in} \qquad (3\text{-}23)$$

The peak collector current of each transistor is given by

$$I_C = \frac{I_L}{n} + I_{mag} \qquad (3\text{-}24)$$

Assuming that $I_{mag} \ll I_L/n$, then

$$I_C = \frac{I_L}{n} \qquad (3\text{-}25)$$

The following expression is derived for the transistor collector working current in terms of output power, efficiency, and duty cycle, as shown in Sec. 3-2-1:

$$I_C = \frac{P_{out}}{\eta \delta_{max} V_{in}} \qquad (3\text{-}26)$$

Assuming a converter efficiency $\eta = 80$ percent and $\delta_{max} = 0.8$, then the working transistor collector current is

$$I_C = \frac{1.6 P_{out}}{V_{in}} \qquad (3\text{-}27)$$

3-3-3 Limitations of the Push-Pull Circuit

Although the push-pull converter offers some advantages, such as nonisolated base drives and simpler driver circuitry, it also presents a number of disadvantages which make its use as an off-the-line switcher almost not practical.

The first limitation is the voltage rating of the transistors, which should handle twice the input voltage to the converter plus any leakage spike that might result because of transformer leakage inductance as shown in Fig. 3-11. This means that switching transistors able to block over 800 V (for use in 230-V ac lines) must be specified for an off-the-line push-pull converter. This may present a problem for high-power converters, since high-current–high-voltage transistors are not all that common and also tend to be expensive.

Figure 3-11 also shows the second and more severe of the problems associated with push-pull circuits, that is, transformer core saturation. In today's switching power supplies, ferrite core material is widely used because of its low losses at high frequencies of 20 kHz and above. Unfortunately the ferrites also have a high susceptibility to core saturation because of their low flux density, which is usually around 3000 gauss (G). Therefore, a small amount of dc bias in the core will drive it to saturation. This is exactly what happens with the push-pull circuit. When one transistor switches on, the flux swings in one direction of the *B-H* curve in order to reverse direction when the first transistor switches off and the second transistor switches on. In order for the two areas of the flux density to be equal, the saturation and switching characteristics of the switching transistors must be identical under

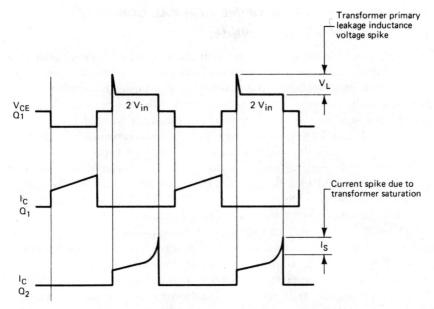

FIGURE 3-11 These are practical voltage and current waveforms associated with the push-pull converter shown in Fig. 3-10.

all working conditions and temperatures. If the transistor characteristics are not identical, "flux walking" to one direction of the B-H curve occurs, driving the core into the saturating region. Saturation of the core means high collector current spike in one of the transistors, as shown in Fig. 3-11.

This excessive current produces large amounts of power loss in the transistor, heating it up, which in turn further unbalances the transistor characteristics, thus saturating the core even more, producing even higher saturation currents, and so on. This vicious cycle will continue until the transistor goes into a thermal runaway, which leads to destruction.

Two solutions to the problem are possible. First is gapping the core, which means an increase in leakage inductance and therefore the need of a dissipative snubber at the expense of lower efficiency. Second is using a symmetry correction circuit, which ensures balance operation of the power transformer by modifying and keeping the on-off ratios of the driver generator equal. Unfortunately this method requires extra circuitry, which adds to the cost and complexity of the converter.

The disadvantages of the push-pull circuit may be alleviated by using the half-bridge or full-bridge power converter. The half-bridge converter is very popular among power supply designers, and it is extensively discussed in Sec. 3-4-1.

3-4 CIRCUIT VARIATIONS OF THE PUSH-PULL CONVERTER

3-4-1 The Half-Bridge Converter

As mentioned previously, there are two main reasons for developing the half-bridge circuit. One is to be able to work the converter from both 115- and 230-V ac input voltages without worrying about using high-voltage transistors, and the second is either to incorporate a simple means of balancing the volt-seconds interval of each switching transistor without gapping the power transformer or to use expensive symmetry correction circuits. Figure 3-12 shows the basic dual input voltage half-bridge converter.

Notice that in the half-bridge configuration the power transformer has one side connected to a floating voltage potential created by the series capacitors C_1 and C_2 which has a value of $V_{in}/2$, 160 V dc at nominal input voltage. The other end of the transformer is connected at the junction of the Q_1 emitter and Q_2 collector through a series capacitor C_3. When Q_1 turns on, this end of the transformer goes to the positive bus, generating a voltage pulse of 160 V. When Q_1 turns off and Q_2 turns on, the polarity of the transformer primary reverses, since it is now connected to the negative bus, generating a negative pulse of 160 V. The turn-on–turn-off action of Q_1 and Q_2 therefore will generate a 320-V peak-to-peak square wave, which in turn is stepped down, rectified, and filtered to produce the output dc voltage.

We have therefore succeeded with this converter topology in achieving the first goal of reducing the voltage stress imposed on the switching tran-

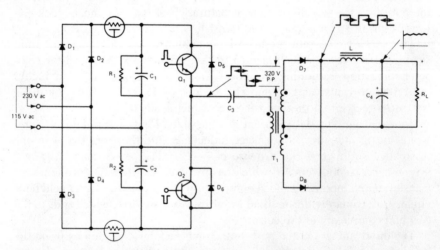

FIGURE 3-12 Basic half-bridge realization. Notice that the same transistors and transformer are used for both 115- and 230-V ac inputs. D_5 and D_6 are leakage inductance commutating diodes.

sistors to no more than V_{in}. Now lower voltage transistors may be specified, and in practice 400-V transistors are commonly used in this application.

There is one minor price to pay when using the half-bridge circuit, that is, since the transformer voltage has been reduced to $V_{in}/2$, the transistor working current will double. If we assume a converter efficiency of 80 percent and a duty cycle $\delta_{max} = 0.8$, then the transistor working current is

$$I_C \approx \frac{3P_{out}}{V_{in}} \qquad (3\text{-}28)$$

The second objective of this topology is to balance the volt-second integral of each switching transistor automatically in order to avoid core saturation. Figure 3-12 shows how this is done by inserting a capacitor in series with the transformer primary. Referring to Fig. 3-12, assume that the two switching power transistors have unmatched switching characteristics; that is, transistor Q_1 has a slow turn-off while transistor Q_2 has a fast turn-off.

Figure 3-13a shows the effect of the slow turn-off of transistor Q_1 upon the ac voltage waveform at the junction of Q_1 and Q_2. A volt-seconds unbalance, depicted by the cross-hatched area, is added to one side of the ac voltage waveform. If this unbalance waveform is allowed to drive the power transformer, flux walking will occur, resulting in core saturation, producing transistor collector current spiking, which will lower converter efficiency and may also drive the transistor into thermal runaway to destruction.

By inserting a coupling capacitor in series with the primary transformer winding, a dc bias proportional to the volt-seconds unbalance is picked up by this capacitor, shifting the dc level as shown in Fig. 3-13b, thus balancing the volt-seconds integral of the two switching periods.

A way of decreasing transistor turn-off time is to use Baker clamp diodes in its base circuit, which in effect do not allow the transistor to saturate fully, thus decreasing the storage time. We will discuss more about Baker clamps and their applications in Chap. 4.

3-4-2 The Series Coupling Capacitor

The power transformer coupling capacitor described previously is normally a film type nonpolar capacitor, capable of handling the full primary current. To minimize heating effects a low ESR capacitor must be used or a bank of capacitors may be placed in parallel to lower their ESR and also get the desired capacitance. The following analysis is a guide in selecting the proper value for the coupling capacitor.

By inspecting Fig. 3-12, it can be stated that the coupling capacitor and the output filter inductor form a series resonant circuit. The resonant fre-

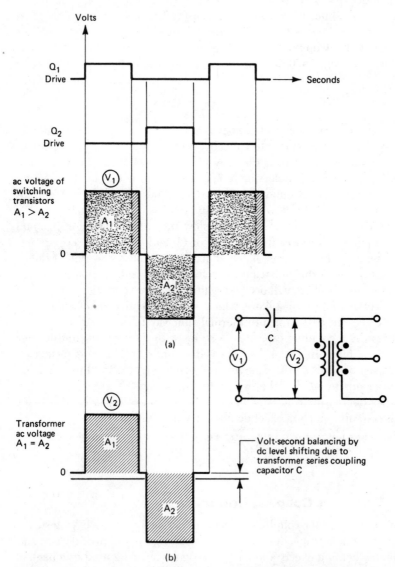

FIGURE 3-13 (a) Waveforms show a volt-seconds unbalance, depicted by the cross-hatched area, on the ac voltage before the series capacitor. This unbalance is due to slow turn-off of transistor Q_1. (b) The same ac waveform is shown after the series capacitor has shifted the dc level to balance the volt-seconds integral.

quency, from simple circuit theory, is

$$f_R = \frac{1}{2\pi\sqrt{L_R C}} \qquad (3\text{-}29)$$

where f_R = resonant frequency, kHz
C = coupling capacitance, μF
L_R = reflected filter inductance, μH

The reflected filter inductance to the transformer primary is

$$L_R = \left(\frac{N_P}{N_S}\right)^2 L \qquad (3\text{-}30)$$

where N_P/N_S is the transformer primary-to-secondary turns ratio, and L is the output inductance (μH).

Substituting Eq. 3-30 into Eq. 3-29 and solving for C we get

$$C = \frac{1}{4\pi^2 f_R^2 (N_P/N_S)^2 L} \qquad (3\text{-}31)$$

In order for the charging of the coupling capacitor to be linear, the resonant frequency must be chosen to be well below the converter switching frequency. For practical purposes, we will assume that the resonant frequency must be about one-fourth the switching frequency, expressed as

$$f_R = 0.25 f_S \qquad (3\text{-}32)$$

where f_S is the converter switching frequency (kHz).

EXAMPLE 3-1

Find the coupling capacitance of a converter working at 20 kHz, which has an output inductance of 20 μH and a transformer turns ratio of 10.

SOLUTION

$f_V = \frac{1}{4} f_S$

Since the switching frequency is 20 kHz, the resonant frequency from Eq. 3-32 will be f_R = 5 kHz. The reflected inductance from Eq. 3-30 is L_R = $10^2(20 \times 10^{-6})$ = 2000×10^{-6} = 2 mH. Therefore the coupling capacitor is

$$C = \frac{1}{4(3.14)^2(25 \times 10^6)(2 \times 10^{-3})} = 0.50 \ \mu\text{F}$$

Another important aspect relating to the value of the coupling capacitor is its charging voltage. Since this capacitor charges and discharges every half cycle and shifts the dc level as shown in Fig. 3-12, its voltage either adds or subtracts to the $V_{in}/2$ value impressed on the transformer primary. Of course the most critical design criterion occurs when the voltage of the charging capacitor bucks the $V_{in}/2$ value, since if this voltage is excessive, it interferes with the converter regulation at low line.

There are two steps by which to check this voltage and in turn to correct the calculated capacitance. The capacitor charging voltage is given by

$$V_C = \frac{I}{C} \, dt \qquad (3\text{-}33)$$

where I = average primary current, A
C = coupling capacitance, μF
dt = time interval capacitor is charging, μs

The time interval by which the capacitor is charging is given by

$$dt = \frac{T}{2} \, \delta_{max} \qquad (3\text{-}34)$$

and

$$T = \frac{1}{f_s} \qquad (3\text{-}35)$$

where T = switching period, μs
δ_{max} = duty cycle
f_s = switching frequency, kHz

For a 20-kHz converter having a duty cycle of 0.8 (80 percent), the charging interval will be 20 μs.

The charging voltage V_C must have a reasonable value anywhere between 10 and 20 percent of $V_{in}/2$; that is, if $V_{in}/2 = 160$ V nominal, then $16 \leq V_C \leq 32$ V for good converter regulation. If the charging voltage exceeds these limits, a new calculation is needed to get a better capacitance value. This value then will be given as

$$C = I \, \frac{dt}{dV_C} \qquad (3\text{-}36)$$

where I = average primary current, A
dt = charging interval, μs
dV_C = arbitrary number from 16 to 32 V

The arbitrary number for dV_C may be so chosen to give a capacitance value close to a real-world standard value. Substituting the calculated capacitance given by Eq. 3-36 we can get the voltage rating of the series coupling capacitor. Although this theoretical voltage rating may be quite low, in practical designs film capacitors with 200-V ratings are normally used.

EXAMPLE 3-2

Suppose we were to use the capacitor calculated in Example 3-1 in a 200-W, 20-kHz half-bridge converter. Verify if the calculated value of 0.50 μF is acceptable. If not, then calculate the appropriate value of the coupling capacitor.

SOLUTION

From Eq. 3-28 we calculate the transistor working current at nominal voltage to be

$$I_C = \frac{3(200)}{320} = 1.86 \text{ A}$$

Assuming that the converter has an input voltage tolerance of ± 20 percent, then the heaviest current of the transistor will occur at low line. Making the correction, the worst-case collector current will be

$$I_C = 1.85 + 0.2(1.86) = 2.3 \text{ A}$$

Using Eq. 3-33 the coupling capacitor charging voltage is

$$V_C = \frac{2.3(20 \times 10^{-6})}{0.5 \times 10^{-6}} = 90 \text{ V}$$

A charging voltage of 90 V is too high and will interfere with converter regulation at low line; therefore a new value of coupling capacitance must be found. Choosing a charging voltage of 30 V and using Eq. 3-36 we get

$$C = \frac{2.3(20 \times 20^{-6})}{30} = 1.5 \text{ } \mu\text{F}$$

A standard value of 1.5 μF may be used with a minimum voltage rating of 30 V. Practical capacitor voltage ratings of 200 V are more commonly used for safety purposes.

3-4-3 The Commutating Diodes

In Fig. 3-12, showing the basic half-bridge converter, diodes D_5 and D_6 were used across transistors Q_1 and Q_2, respectively. These diodes are called *commutating diodes* and they have a dual function.

1. When the transistor turns off, the commutating diode steers transformer leakage inductance energy back to the main dc bus. Thus high-energy leakage inductance spikes, such as the ones shown in the V_{CE} waveform of Fig. 3-11 associated with the push-pull circuit, are not present.

2. The commutating diode prevents the collector of the on transistor from swinging negative in respect to its emitter in the event of a sudden off-load situation due to an increase in transformer flux. In such an event the commutating diode will bypass the transistor until the collector goes positive again, preventing the device from an inverse conduction and possible damage.

The commutating diodes must be fast-recovery types with a blocking voltage capability of at least twice the transistor collector-to-emitter off voltage. In practice, diodes with a reverse blocking voltage of 450 V are commonly used.

3-5 THE FULL-BRIDGE CIRCUIT

During our discussion of the half-bridge circuit we noted that, although we successfully reduced the voltage stress of the switching transistors at turn-off to half the input dc voltage, the price to pay was doubling the collector current at turn-on, as compared to the push-pull circuit. This constraint is all right for low- and medium-power applications, but it becomes more prohibitive for high-power applications where high-voltage–high-current transistors are less available.

A way of getting around this problem is to construct a circuit which retains the voltage properties of the half-bridge and the current properties of the push-pull. This circuit is the full-bridge topology and it is shown in Fig. 3-14. In this circuit diagonally opposite transistors Q_1 and Q_4 or Q_3 and Q_2 are simultaneously turned on.

This transistor action causes the voltage imposed on the transformer primary to swing between $+V_{in}$ and $-V_{in}$. Thus the transistors never see a collector off-voltage of more than V_{in}. Also the current through them is half of an equivalent half-bridge circuit.

A drawback of this topology is the necessity of using four transistors, and since diagonally opposite transistors are on at the same time, isolated base drives for each transistor must be used.

Assuming a converter efficiency of 80 percent and a 0.8 duty cycle, then the transistor working current is

$$I_C = \frac{1.6P_{out}}{V_{in}} \tag{3-37}$$

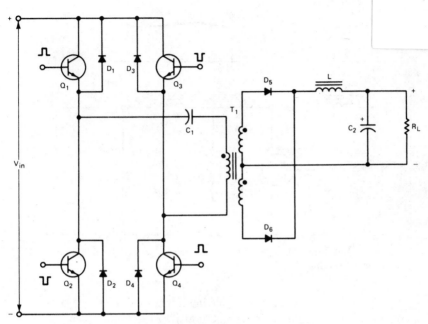

FIGURE 3-14 The full-bridge converter circuit.

All other properties of this converter remain the same as for the half-bridge converter, and all the formulas developed for the calculation of other elements may be used here as well.

3-6 A NEW ZERO OUTPUT RIPPLE CONVERTER

All the circuit topologies discussed in the previous sections produce an output current which has a certain amount of ripple. In recent years, a new converter topology has emerged called the Ćuk converter, named after its inventor, Dr. S. Ćuk, which with appropriate transformer design can have an output ripple of zero.

The basic nonisolated converter is shown in Fig. 3-15. Operation of the circuit is as follows. When transistor Q_1 is off, diode D_1 is switched on, charging capacitor C_1 by the input current I_1. During the interval when Q_1 switches on, D_1 turns off, grounding the positive terminal of capacitor C_1. Thus a current I_2 flows through L_2, producing a negative output voltage at the load.

Since this converter has the properties of a combined buck-boost topology, and since the energy transfer is capacitive, the input and output currents

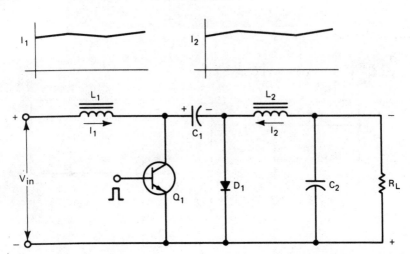

FIGURE 3-15 Realization of the basic Ćuk converter and its associated input and output current waveforms.

are very close to true dc quantities having negligible switching ripple. But "negligible switching ripple" does not actually mean "no ripple," which is the ultimate goal of this topology. Fortunately, the zero output ripple topology may be achieved by making the following observations. In order for the average dc voltage across each inductor to be zero, the two waveforms must be equal and identical. But in order to achieve this the two inductors must share the same core and they must have the same number of turns. Figure 3-16 shows the implementation of this idea.

Since the two coupled inductors now form a transformer, the effective inductance of each winding is altered by the alternating inductive energy transfer. If we were to set a turns ratio of 1:1, both inductances would double, thus reducing the input and output ripple to half those of the uncoupled converter. This is indeed a very important observation, since if we altered the turns ratio to the point where the primary-to-secondary turns ratio matches the inductive coupling coefficient of the transformer, the output current ripple could be completely eliminated, as shown in Fig. 3-16.

Although this is a very useful circuit, its applications are limited to designs where no input-to-output isolation is required. For use of this converter in an off-the-line configuration, isolation of input and output is a must. The following discussion shows how isolation may be achieved in the Ćuk converter, and the three basic steps in achieving this are shown in Fig. 3-17.

The first step is to split the coupling capacitance C_1 into two series capacitors C_A and C_B, as shown in Fig. 3-17a. Since the average dc voltage at the connection point of the two capacitors is indeterminate and floating, it may be forced to zero by placing inductor L between this point and ground,

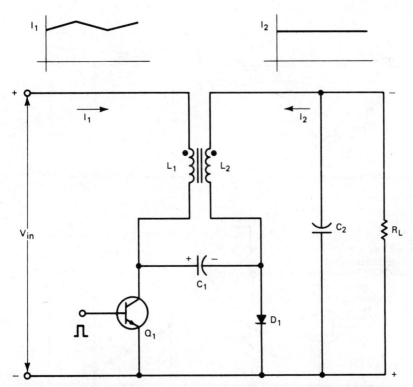

FIGURE 3-16 The coupled inductor version of the Ćuk converter and its associated current waveforms.

as shown in Fig. 3-17*b*. If we make this inductance very large, the current it diverts from the two series capacitors is negligible and therefore the operation of the converter remains unaffected.

Now by changing inductor *L* into an isolation transformer, the desired dc isolation has been achieved. The basic isolated version of the Ćuk converter is shown in Fig. 3-17*c*. As we mentioned, the isolated version of the Ćuk converter retains all the properties of the nonisolated converter, and therefore, the zero ripple extension may be used successfully in this case also.

Figure 3-18 shows the coupled inductor zero output ripple dc isolated Ćuk converter and its waveforms. Notice that in this circuit the transfer capacitors C_A and C_B have been inserted in the other side of their respective transformer windings. This change does not affect the operation of the converter.

Although coupling the input and output inductors reduces the output ripple, it also introduces an undesirable side effect, that is, an output polarity

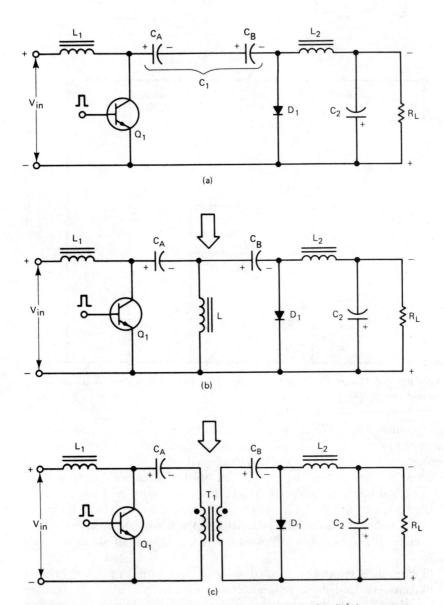

FIGURE 3-17 The three basic steps of transforming a nonisolated Ćuk converter to an isolated one. (*a*) Splitting the coupling capacitance C_1 into series capacitors C_A and C_B; (*b*) placing inductor L between point and ground; (*c*) changing inductor L into an isolation transformer.

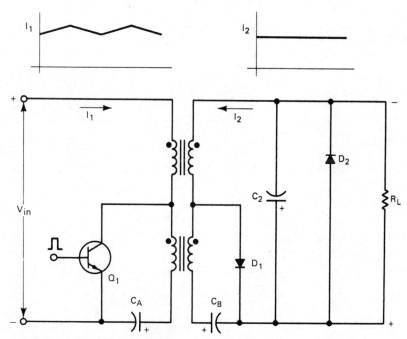

FIGURE 3-18 Coupled inductor zero ripple dc isolated Ćuk converter. Diode D_2 is the output clamp diode, holding the output to a diode drop due to polarity reversal at start-up.

reversal at start-up. Although the reverse polarity pulse is of short duration, it may prove fatal to sensitive electronic loads. The addition of a clamp diode D_2, as shown in Fig. 3-18, limits the transient pulse to a volt or less, thus protecting sensitive loads.

At the writing of this book, the Ćuk converter and its associated applications are protected by a series of patents, and its use in commercial or other designs may require a license agreement with its inventor.

3-7 THE BLOCKING OSCILLATOR OR RINGING CHOKE CONVERTER

The fixed frequency pulse-width modulated (PWM) power supply design has been exclusively used in modern designs, mainly due to the availability of PWM control circuits. On the other hand, for many low-power (approximately 10 to 50 W), inexpensive power supplies, the variable frequency blocking oscillator converter circuit has been successfully implemented in many designs. Figure 3-19 depicts a practical realization of a low-power variable frequency blocking oscillator circuit, with its associated waveforms. Operation of the circuit is as follows.

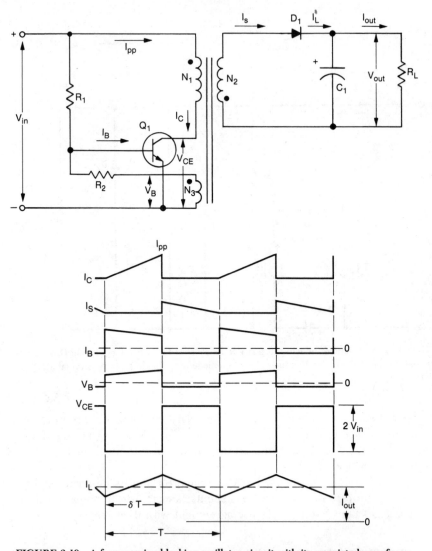

FIGURE 3-19 A free running blocking oscillator circuit with its associated waveforms.

At initial power-up transistor Q_1 receives a certain amount of base current through resistor R_1, causing the transistor to saturate. Consequently, a peak current I_{PP} flows in the primary winding of transformer T_1 equal to

$$I_{PP} = I_C = \frac{V_{in}}{L_P}(\delta_{max}T) = \frac{V_{in}}{L_P}t_{on} \qquad (3\text{-}38)$$

where $(\delta_{max}T)$ is the maximum transistor on time t_{on}.

At the same time, flux is building up on winding N_3 developing a voltage potential V_B. Therefore base current to keep the transistor in saturation is now supplied through resistor R_2. The magnitude of the base current is

$$I_B = \frac{V_B}{R_2} = \left(\frac{N_3}{N_1}\right)\left(\frac{V_{in}}{R_2}\right) \tag{3-39}$$

Since in a transistor the collector current is a function of the base current and the transistor gain, that is, $I_C = \beta I_B$, then the transformer primary current will increase up to the point where the above product reaches maximum. Beyond this threshold the base circuit cannot sustain any more collector current increase, resulting in transistor Q_1 cutoff and a transfer of power from the transformer primary to the secondary and consequently after rectification by D_1 to the load.

The transistor used in this circuit must be so chosen as to have a minimum collector-to-base (V_{CBO}) rating given by

$$V_C = V_{in} + \left(\frac{N_1}{N_2}\right)V_{out} + (\text{leakage spikes}) \tag{3-40}$$

For off-the-line power supplies, transistors with $V_{CBO} \geq 800$ V are commonly used.

Since the operating frequency of this type of converter is variable, care must be taken to design the circuit so that its lowest operating frequency does not drop below 20 kHz, so that the power supply does not become audible. Remember that the frequency of operation is higher at light loads and lower at full load.

3-7-1 The Blocking Oscillator Transformer

Since this topology is basically a flyback converter, the transformer-choke has to be designed very carefully. To calculate the primary turns of the transformer, a full-load frequency of operation must be chosen by the designer. Then the transistor on time must be picked so that it will run at a preferred 50 percent duty cycle maximum. With these factors established, the primary transformer inductance in microhenries is calculated as follows:

$$L = \frac{V_{in,min}}{I_{PP}} t_{on} \tag{3-41}$$

where $V_{in,min}$ is the minimum input voltage which the transistor will switch, t_{on} is the maximum transistor on time, and I_{PP} is the peak primary current through the transistor.

Once the transformer primary inductance is calculated, the primary number of turns is found by

$$N_P = \frac{L(I_{PP})}{A_e B_{max}} \qquad (3\text{-}42)$$

where A_e is the effective area of the core and B_{max} is the maximum allowable
working flux density of the core.

In order to avoid core saturation, an air gap will be used. The length of the
air gap in ~~millimeters~~ is given by
centimeters

$$l_g = \frac{A_e(N_P)^2 10^{-8}}{0.8L} \qquad (3\text{-}43)$$

A practical way to calculate the secondary windings number of turns is
to determine first the transformer primary turns per volt ratio:

$$r = \frac{N_P}{V_{in,min}} \qquad (3\text{-}44)$$

Since we chose $t_{on} = t_{off}$, then the secondary winding(s) number of turns
will be

$$N_S = r(V_s + V_{ss}) \qquad (3\text{-}45)$$

where V_s is the nominal secondary output voltage desired and V_{ss} is the
voltage drop encountered in the conductors and the output rectifier.

The above calculations will result in a very good first-order approximation,
and minor adjustments to produce exact results may be made by the designer
during actual circuit implementation and testing.

3-7-2 A MOSFET Blocking Oscillator Converter

A practical implementation of a MOSFET blocking oscillator converter using
a single Motorola (or equivalent) power TMOS transistor is shown in Fig.
3-20. This design may well be used for off-the-line switching power supplies,
and its regulation is good enough for many applications.

The operation of the circuit is as follows. At power turn-on, in phase
transformer windings N_1 and N_2 will cause the circuit to oscillate. Oscillations
start when capacitor C_1 is charged through a large resistor R_1 connected to
the supply rail. Resistor R_2 is used to limit Q_2 collector current. The on time
of the oscillation cycle is terminated by transistor Q_2 which senses the ramped
source current of Q_1. Capacitor C_1 is therefore charged on alternate half
cycles by transistor Q_2 and forward biased zener Z_2.

Regulation is provided by taking the rectifier output of sense winding N_3
and applying it as a bias to the base of Q_2 via zener diode Z_1. The collector

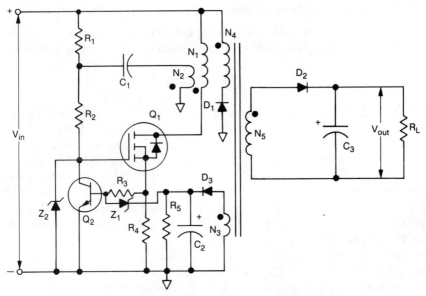

FIGURE 3-20 Practical implementation of a blocking oscillator converter using a power MOSFET. (*Courtesy Motorola Semiconductor Products, Inc.*)

of Q_2 modulates the gate of Q_1, shortening or lengthening its on time, thus keeping the output voltage constant.

3-8 THE SHEPPARD-TAYLOR CONVERTER

The following new converter topology, like the Ćuk converter, has the very desirable features of nonpulsating currents in both input and output ports, thus significantly reducing both conducted and radiated noise. The circuit is best suited for high operating frequency applications and therefore the use of MOSFETs is inherent. The new topology is a boost converter cascaded with a buck converter, and it may work in off-the-line applications from unfiltered dc inputs. The input reservoir capacitor is no longer needed, at the expense of a total lack of output hold-up time. Its inventors David Sheppard and Brian Taylor claim no patent restrictions for the converter, and it is therefore free for unrestricted use. It is the belief of the author of this book that the converter has many merits and features, many yet unexplored, and it is especially suited in today's high-frequency applications (above 100 kHz). Readers, power supply designers, and students involved in converter research are therefore encouraged to explore this topology for better understanding and optimum usage. The name "Sheppard-Taylor converter" has been arbitrarily coined by this author.

3-8-1 Circuit Analysis of the Sheppard-Taylor Converter

The basic circuit schematic of the Sheppard-Taylor converter is presented in Fig. 3-21, in its nonisolated version. Operation of the circuit is as follows. Assume that MOSFETs Q_1 and Q_2 are off. Then an input current I_{in} will flow through L_1, D_1, C, and D_2 to the negative rail. This current flow will charge the nonpolarized type capacitor C, so that the plate connected to the cathode of D_1 is at an initial potential of approximately V_{in}, while the opposite plate is clamped by D_2 to the negative rail. No current is flowing to the output of the converter due to D_3. Note that as C continues to charge, the current in L_1 diminishes linearly. When Q_1 and Q_2 are switched on, diodes D_1 and D_2 will be reverse-biased, prohibiting any further current flow from the input terminals. At that instant, capacitor C is in essence connected across the output terminals of the converter, and because of its charged polarity, D_3 will be forward-biased, allowing output current to flow through L_2, D_3, Q_1, C, and Q_2.

By inspection we can see that the voltage across L_1 is proportional to the sum of V_{in} and V_C, the voltage across capacitor C when Q_1 and Q_2 are off. The moment Q_1 and Q_2 are switched on, the diminishing current in L_1 changes and begins to ramp up, storing new energy in the inductor L_1. During the next cycle when Q_1 and Q_2 are once again turned off, capacitor C will start charging, from the energy stored in L_1, through diodes D_1 and D_2. In attempting to balance its volt-second products, L_1 will cause C to change to some new voltage, which is inversely proportional to the conduction duty cycle of Q_1 and Q_2. Meanwhile D_3 is again reverse-biased, and

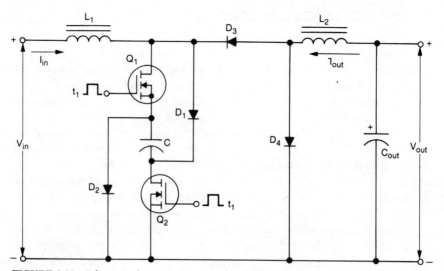

FIGURE 3-21 Schematic diagram of a nonisolated Sheppard-Taylor converter.

output current is now sustained through the free-wheel diode D_4. In order to derive certain equations to facilitate components rating and selection and relate input/output parameters, the circuits of Fig. 3-22 will be used.

From Fig. 3-22, Q_1 and Q_2 are on during period t_p, and off during period t_d. The duty cycle δ is

$$\delta = \frac{t_p}{t_p + t_d} \tag{3-46}$$

As previously mentioned, the average value of the voltage across inductor L_1 will be the sum of V_{in} and V_c.

From Fig. 3-22a then, applying Kirchhoff's voltage law, during period t_p,

$$V_{L_1} = V_{in} - (-V_C)$$

or

$$V_{L_1} = V_{in} + V_c \tag{3-47}$$

and

$$\frac{dI_1}{dt_p} = \frac{V_{in} + V_c}{L_1} \tag{3-48}$$

Similarly, during the same period the average value of the voltage across L_2 will be

$$V_{L_2} = V_C - V_{out} \tag{3-49}$$

and

$$\frac{dI_{out}}{dt_p} = \frac{V_c - V_{out}}{L_2} \tag{3-50}$$

From Fig. 3-22c, let I_{inp} and I_{outp} be the peak-to-peak values of I_{in} and I_{out}, respectively, during the on period t_p. Then by integrating Eqs. 3-48 and 3-49, we get

$$I_{inp} = \frac{t_p(V_{in} + V_c)}{L_1} \tag{3-51}$$

and

$$I_{outp} = \frac{t_p(V_c - V_{out})}{L_2} \tag{3-52}$$

During the off period t_d, assuming the peak-to-peak values of I_{in} and I_{out} to be I_{ind} and I_{outd}, respectively, we can show that

$$I_{ind} = \frac{t_d(V_c - V_{in})}{L_1} \tag{3-53}$$

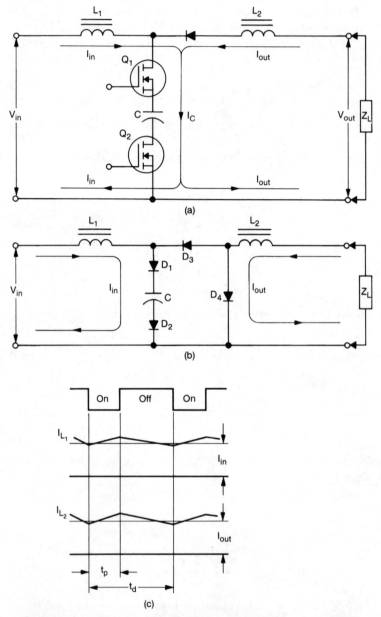

FIGURE 3-22 Circuit analysis of the Sheppard-Taylor converter. (*a*) Current paths when MOSFETs are on, (*b*) current paths when MOSFETs are off, and (*c*) associated output current waveforms during the on and off periods.

and

$$I_{\text{out}d} = \frac{t_d V_{\text{out}}}{L_2} \tag{3-54}$$

If we assume that during the on period t_p, the input voltage V_{in}, and output load R of the converter remain constant, then the charge extracted from C during this period will be proportional to the "droop" in voltage across C. This droop in voltage may be written as ΔV_c, and the charge equation will be

$$(\Delta V_c)C = I_c t_p \tag{3-55}$$

where $I_c = I_{\text{in}} + I_{\text{out}}$.

It can also be shown that

$$I_c = \frac{V_{\text{out}}^2}{R V_{\text{in}}} \tag{3-56}$$

Therefore

$$(\Delta V_c)C = \frac{t_p V_{\text{out}}(V_{\text{out}} + V_{\text{in}})}{R V_{\text{in}}} \tag{3-57}$$

Equally, if we assume during the off period t_d, that V_{in} and R remain constant, then the voltage ramp across C can be regarded as having the same magnitude as the drop in voltage and can therefore be regarded also as ΔV_c. Hence,

$$(\Delta V_c)C = I_c \tag{3-58}$$

and

$$t_d = \frac{t_d V_{\text{out}}^2}{R V_1} \tag{3-59}$$

Since we assumed that V_{in} and R remain constant during both t_p and t_d, it follows that the volt-second products across L_1 are equal. Therefore, equating Eqs. 3-51 and 3-53, we get

$$t_p(V_{\text{in}} + V_c) = t_d(V_c - V_{\text{in}}) \tag{3-60}$$

Solving for V_c,

$$V_c = \frac{V_{\text{in}}}{1 - 2\delta} \tag{3-61}$$

where δ is the duty cycle defined as

$$\delta = \frac{t_p}{t_p + t_d}$$

Similarly, Eqs. 3-52 and 3-54 can be equated to give the classic buck regulator output voltage relationship:

$$V_{out} = V_c \delta \tag{3-62}$$

or

$$V_{out} = \frac{V_{in} \delta}{1 - 2\delta} \tag{3-63}$$

The above equation shows an unusual and unique characteristic of the new converter, that is, the output voltage approaches infinity as the duty cycle approaches 50 percent! This characteristic gives the converter some extremely interesting and unique features, the most important being the ability of the converter to hold output regulation with very low input voltage values, as shown by Eq. 3-63.

The discussion so far has centered around the nonisolated version of the converter. An isolated version of the converter may easily be implemented as shown in Fig. 3-23. In the isolated version, D_3 is replaced by a pair of rectifiers D_{3A} and D_{3B}, and an isolation high-frequency transformer is inserted between this pair of rectifiers.

Let the turns ratio of the transformer be N, where $N = N_p/N_s$. Then Eqs. 3-52, 3-57, and 3-63 should be modified to read, respectively,

$$I_{outp} = \frac{t_p(V_c - NV_{out})}{NL_2} \tag{3-64}$$

$$(\Delta V_c)C = \frac{t_p V_{out}(NV_{out} + V_{in})}{NRV_{in}} \tag{3-65}$$

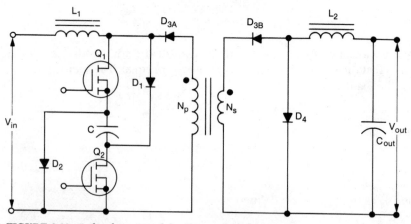

FIGURE 3-23 Isolated version of the Sheppard-Taylor converter.

and

$$V_{out} = \frac{V_{in}}{N(1 - 2\delta)} \qquad (3\text{-}66)$$

3-8-2 Features of the Sheppard-Taylor Converter

As previously mentioned, this converter is ideal for high-frequency applications (above 100 kHz). At these frequencies MOSFETs are highly recommended. Even if bipolars were to be considered, examination of Eq. 3-61 shows clearly that during an overcurrent condition, the conduction time of the power switches is small; therefore, the inherent storage times of bipolars do not provide a control range broad enough to be usable. At higher frequencies the value of the switching capacitor C is also minimized, a desirable effect in reducing both size and weight and also making the value of C practical. We mentioned before that this converter also holds output regulation within a wide input voltage range. This is a very desirable feature particularly for off-the-line switching power supplies enabling the converter to operate worldwide without modification of the input circuitry. Although the converter may, for off-the-line applications, operate from unfiltered dc inputs, the absence of the input reservoir capacitor may not be desired if a minimum output hold-up time is required. A reservoir capacitor can be used to increase hold-up time, and for a given value of capacitance, the output hold-up time will increase by a factor of 2 over conventional topologies.

Since the input and output currents of the converter are nonpulsating as shown in Fig. 3-22c, ac line or dc output filtering to meet local, national, or international noise requirements, such as FCC or VDE, is greatly simplified, or unnecessary.

Snubbers across the MOSFETs and the power transformer are not necessary using this topology, greatly reducing component count and improving efficiency.

Also, perfect synchronization of switching Q_1 and Q_2 on and off is not essential. Lack of synchronization does not lead to malfunction or to failure of any device or component.

3-9 HIGH-FREQUENCY RESONANT CONVERTERS

High-frequency resonant converters, whether parallel or series, have become increasingly popular among power supply designers, especially at frequencies above 100 kHz, because they offer small size, good reliability, and reduced EMI/RFI. Recent advances and price reductions in both control ICs and power MOSFETs have made resonant sine wave converters even more popular. Series resonant topologies are being widely used because of their tolerance to switching transistor transition times and reverse recovery times and their ease of operation. Some of the advantages of the resonant sine wave converter, compared to other conventional topologies, follow: With this converter, there is higher overall efficiency at a given power level, mainly due to the absence of switching losses at the power switch and the rectifiers. Lower losses in turn mean smaller heat sinks, hence reduction in size and weight of the overall package. Because the voltage is switched when the drain current of the MOSFET switching transistors is zero, operation at higher frequencies is possible, resulting in smaller magnetic elements and filter components. Since the currents in the series resonant converter are sinusoidal in nature, the absence of high di/dt current changes, associated with square wave converters, means low levels of EMI/RFI emissions. In the following paragraphs, a series resonant converter will be discussed, and the basic design parameters will be given.

3-9-1 The Basic Sine Wave Series Resonant Converter

A series resonant power stage and its associated waveforms is shown in Fig. 3-24. Operation of the circuit is as follows.

Assume that MOSFETs Q_1 and Q_2 are off and that capacitor C_R is completely discharged. Drive pulse t_1 is applied at the gate of Q_1, turning it on. Current i_{Q_1} is thus flowing through resonant inductor L_R, the primary winding of transformer T_1, and the resonant capacitor C_R. Since the voltage across the transformer primary is fixed by its turns ratio N and the output voltage, and because the resonant current i_{Q_1} is controlled by the series resonant network $L_R C_R$, it increases in a sinusoidal manner starting from zero, charging resonant capacitor C_R and delivering energy through transformer T_1 to the load. When the peak of the sine wave current is reached, capacitor C_R is clamped by diode D_1 to the positive rail. At that point the voltage across inductor L_R ceases to increase, and its energy is released through T_1 to the load causing the current i_{Q_1} to decrease toward zero. Transistor Q_1 remains on until the current in the resonant network reaches zero, at which point Q_2 is turned on and the above cycle repeats itself but in the opposite direction, since the energy is now drawn from the previously charged resonant capacitor C_R. Thus the composite resonant sinusoidal current i_R is developed, as shown in Fig. 3-24, which produces a proportional sinusoidal primary voltage. The output of the center-tapped transformer secondary voltage is then rectified and filtered by capacitor C_{out} to produce the dc output voltage V_{out}.

Note that the leakage inductance of the transformer is in series with the resonant inductor, and it should be taken into account when designing the power stage.

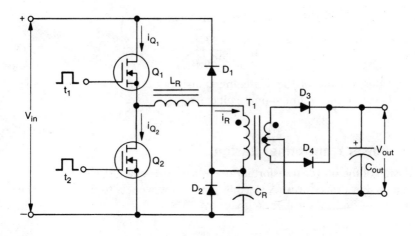

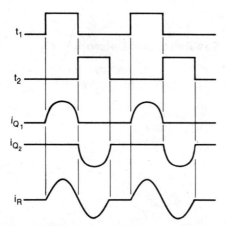

FIGURE 3-24 A high-frequency sine wave series resonant converter and its associated waveforms.

3-9-2 Transistor Selection in Series Resonant Converter

The maximum current in a series resonant converter occurs at high input line. Since this current also flows through the switching transistors, MOSFETS Q_1 and Q_2 must be selected to handle a minimum current of

$$I_{\max} = \left(2\,\frac{V_{\text{in,max}}}{V_{\text{in,min}}} - 1\right)I_{\text{low}} \tag{3-67}$$

where I_{low} is the peak current at low line, given by

$$I_{\text{low}} = \left(\frac{\pi}{2}\right)I_{\text{pri}} \tag{3-68}$$

and I_{pri} is the transformer primary current given by

$$I_{pri} = \frac{2P_{in}}{V_{in,min}} \tag{3-69}$$

The blocking voltage of the switching transistors used must have a value greater than the maximum supply voltage $V_{in,max}$.

3-9-3 Power Transformer Design

The design of the power transformer in the series resonant converter follows classical design procedures. A starting point, however, is to determine the primary to secondary turns ratio given by

$$N = \frac{V_{in,min}}{2V_{out}} \tag{3-70}$$

3-9-4 Design of the Series Resonant Network $L_R C_R$

The characteristic impedance of the series resonant network as a function of input voltage and output power is given by the following equation:

$$Z_{out} = \frac{\eta V_{in,min}^2}{2\pi P_{out}} \tag{3-71}$$

where η is the efficiency of the converter.

Then the resonant capacitor is

$$C_R = \frac{1}{2\pi f Z_{out}} \tag{3-72}$$

where f is the frequency of operation of the converter.

The resonant inductor is

$$L_R = \frac{Z_{out}}{2\pi f} \tag{3-73}$$

3-9-5 Design of the Resonant Inductor

First calculate the maximum circuit energy storage, using Eq. 3-74:

$$W_{max} = \frac{1}{2} L_R I_{max}^2 \tag{3-74}$$

Calculate the energy storage capability of the core given by

$$H1_e = \frac{2W_{max}10^8}{BA_e} \tag{3-75}$$

where B is the chosen core operating flux density in gauss. (A good starting point is $B = B_{sat}/2$, and A_e is the core effective area in square centimeters.)

Since

$$H1_e = NI_{max} \tag{3-76}$$

solving for N, we get the required resonant inductor number of turns.

$$N = \frac{H1_e}{I_{max}} \tag{3-77}$$

In order to prevent inductor core saturation, an air gap must be introduced in the magnetic path. The value of the air gap is given by

$$1_g = \frac{NI_{max}}{H} = \frac{\mu NI_{max}}{B} = \frac{NI_{max}}{(B/\mu)} \tag{3-78}$$

where μ is the permeability of the chosen core. The gap length is expressed in millimeters.

EXAMPLE 3-3

Calculate the series resonant network $L_R C_R$ of a 200-W converter operating at 200 kHz. Assume an efficiency of 80 percent and an input voltage range of 90 to 130 V ac.

SOLUTION

Solving for minimum and maximum dc input voltage,

$$V_{in,min} = 1.4 \times 90 = 126 \text{ V dc}$$

$$V_{in,max} = 1.4 \times 130 = 182 \text{ V dc}$$

Since the output power is given as $P_{out} = 200$ W and efficiency $\eta = 0.8$, then $P_{in} = 200/0.8 = 250$ W.

Average primary current from Eq. 3-69 is

$$I_{pri} = \frac{2P_{in}}{V_{in,min}} = \frac{2(250)}{126} = 4 \text{ A}$$

Therefore

$$I_{low} = \left(\frac{\pi}{2}\right)I_{pri} = \left(\frac{3.14}{2}\right)4 = 6.28 \text{ A}$$

and

$$I_{max} = \left(2\frac{V_{in,max}}{V_{in,max}} - 1\right)I_{low} = \left(2\frac{182}{126} - 1\right)6.28 = 11.86 \text{ A}$$

Hence the MOSFETs selected must handle a minimum of 11.86 A of drain current.

From Eq. 3.71 we get

$$Z_{out} = \frac{\eta V_{in,min}^2}{2\pi P_{out}} = \frac{0.8(126)^2}{6.28(250)} = 8.09 \ \Omega$$

Hence

$$C_R = \frac{1}{2\pi f Z_{out}} = \frac{1}{6.28(200 \times 10^3)(8.09)} = 0.1 \ \mu F$$

and

$$L_R = \frac{Z_{out}}{2\pi f} = \frac{8.09}{6.28(200 \times 10^3)} = 6.44 \ \mu H$$

Since the maximum energy storage in the inductor occurs at high line, the maximum circuit energy storage required is calculated from Eq. 3-74,

$$W_{max} = \frac{1}{2} L_R I_{max}^2 = \frac{1}{2}(6.44 \times 10^{-6})(11.86)^2 = 453 \ \mu J$$

From Eq. 3-75, and using a ferrite core at 1550 G having an $A_e = 0.9$,

$$H1_e = \frac{2W_{max}10^8}{BA_e} = \frac{2(453)10^{-6}10^8}{1500(0.9)} = 67 \text{ AT}$$

Using Eq. 3-77, the required resonant inductor turns may be calculated as follows:

$$N = \frac{H1_e}{I_{max}} = \frac{67}{11.86} = 6 \text{ turns}$$

The inductor gap required, using Eq. 3-78, is

$$1_g = \frac{NI_{max}}{(B/\mu)} = \frac{67}{[0.15/(4\pi)10^{-7}]} = 0.561 \text{ mm}$$

Since the series resonant converter is operational at high frequencies, extra care must be exercised in selecting components. Litz wire is recommended for the inductor and power transformer to minimize skin effects. Capacitors must be selected for low ESR and ESL and good ripple current ratings. Polypropylene-type capacitors are a good choice at these frequencies.

On the other hand, the output rectifier diodes need not be extremely fast, compared to other converter topologies, due to low di/dt during diode turn-off, inherent in the resonant converter topologies.

3-10 CURRENT-MODE REGULATED CONVERTERS

The current-mode regulated converter differs significantly from conventional PWM converters. Current-mode converters use an inner loop to directly control peak inductor current with the error signal, rather than controlling duty ratio of the pulse-width modulator. Figure 3-25 shows the block diagram of a fixed frequency current-mode forward converter.

As shown, the error amplifier compares the output to a fixed reference voltage, and the resulting error signal V_e is controlling peak switch current, which is proportional to the average output inductor current.

There are several inherent advantages to the current-mode control converter when compared to the conventional PWM converter. These advantages are as follows:

1. Automatic feed-forward improves line regulation.

2. Automatic symmetry correction, due to peak current sensing, makes this scheme suitable to push-pull converters, without the need for complex balance schemes.

3. Automatic current-limiting control, due to the sampling technique used.

4. Simple loop compensation.

5. Improved transient response.

6. Outputs of more than one converter may be easily paralleled while maintaining equal current sharing.

Of course, there are also certain limitations to this scheme, the most important being the following:

1. Loop instability above 50 percent duty cycle.

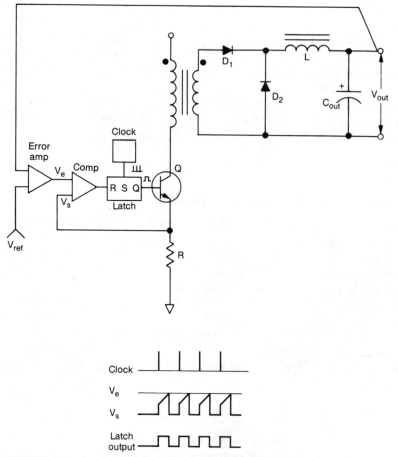

FIGURE 3-25 A current-mode forward converter and its associated waveforms.

2. Less than ideal loop response, caused by peak instead of average inductor current sensing.

3. Tendency toward subharmonic oscillation.

4. Noise sensitivity, particularly at very small inductor ripple current.

However, with careful design the above problems may be reduced or eliminated, making the use of the current-mode control technique a useful and attractive alternative for high-frequency switching power supply design.

Unitrode Corporation is marketing the UC1846 and UC1847 families of current-mode PWM-integrated controllers, which offer all the necessary functions to design high-performance, low-cost power supplies, at working frequencies up to 500 kHz. Chapter 7 describes the UC1846 current-mode control IC.

Two other Unitrode IC controller families—the UC1823 and UC1825—may be configured either in a voltage or current-mode topology, with operating frequencies up to 2 MHz. The reader is referred to Unitrode Corporation's "Linear Integrated Circuits Databook" for specification details and application information for these versatile PWM-integrated circuits.

3-11 THE WARD CONVERTER

The Ward converter is a new dc-dc converter, and the description which follows is the first to appear in any literature ever. The converter was invented by Dr. Michael Vlahos Ward, president of Combustion Electromagnetics, Inc., Arlington, Massachusetts, and a patent is pending on the circuit. Although basic research has been completed by the inventor and a practical working prototype for a specific application has been developed, this converter may have many desirable features still unexplored. It is the belief of the author that this converter has many merits and that it is well suited for high-frequency operation. The fundamental reason that it is suitable for high-frequency operation is that the converter is a member of the family of the so-called zero-current-switching converters. This means that its single controlled switch is turned on at zero current and is turned off when the current in it (or in an external antiparallel diode) flows in the reverse direction. In the Ward converter, as in any other zero-current-switching converter, the switching losses are very small, and the circuit can probably operate efficiently up to several megahertz. Further understanding and research on this topology are therefore encouraged. Figure 3-26 shows the converter and its associated waveforms.

The converter operates by storing the input energy sequentially, first inductively and then capacitively; inductively the time that energy is being delivered to the output circuit, and capacitively for the remaining period. With reference to Fig. 3-26, when the MOSFET transistor switch Q is closed, energy storage inductor L_1 begins to build up current I_1. At the same time capacitor C_1, which was fully charged under steady-state conditions, discharges its energy to produce the first half-sinusoidal input current I_1, and by transformer action this current is transferred to the output, and through rectifier D_2 it charges the output capacitor C_{out}. When switch Q is opened, inductive current I_1 is steered to capacitor C_1 to charge up the capacitor, in preparation for the next cycle when most of the energy is delivered to the output capacitor. In this second period, capacitor C_1 completes the second half of its discharge cycle with the current I_{ii} flowing through diode D_1 across the MOSFET switch, which is then followed by a period when diode D_1 is back biased and voltage V_1 recovers as shown in the last waveform.

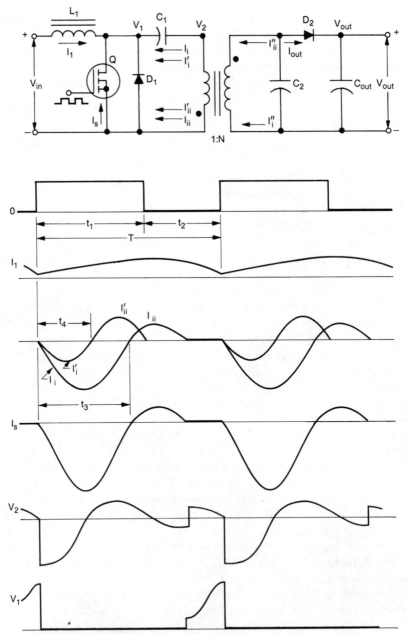

FIGURE 3-26 The Ward converter and its associated waveforms.

Inspection of the waveforms shows that during the energy storage phase when Q is on, a negative charge is placed on the output capacitor C_2, which must be dealt with. The proper handling of this "wrong polarity" charge was the key to the final success of this converter. A precise phasing relationship was determined so that during the energy transfer phase, this negative charge is fed back into the input circuit so as to transfer its charge to capacitor C_1 while simultaneously helping define zero (negative) current at the instant the switch Q is opened. In this way, even under conditions where all the energy on C_1 is transferred to output capacitor C_{out} and almost no reverse current I_{ii} exists to cancel current I_1, primary current I'_{ii}, which is the transformed secondary current I''_{ii}, will guarantee zero current crossing during MOSFET transistor turn-off.

3-11-1 Circuit Analysis of the Ward Converter

There are some simple relationships which define the circuit parameters making up the Ward converter.

The relationship between the input capacitor C_1 and output capacitor C_{out} is somewhat arbitrary and is determined by the requirements of the system and is related to the transformer turns ratio N as follows:

$$\lambda = \frac{C_{\text{out}}N^2}{C_1} \tag{3-79}$$

The parameter λ typically represents the number of cycles of operation of the converter required to fully charge C_{out}. Clearly, the larger λ is, the higher the frequency of operation of the converter and the less sensitive the converter is to a shorted output.

If an output voltage V_{out} is required, then the ideal turns ratio of the transformer will be

$$N = \frac{2V_{\text{out}}}{V_1} \tag{3.80}$$

where V_1 is the voltage to which the energy storage capacitor C_1 is charged at the end of the charge cycle (at time t_2), and whose amplitude is given by

$$V_1 = V_{\text{in}}\left[1 + \sqrt{1 + \left(\frac{I_{L_1}}{I_m}\right)^2}\right] \tag{3-81}$$

where I_{L_1} is the maximum charging current through inductor L_1 at the instant MOSFET Q is turned off, given by

$$I_{L_1} = I_{\text{dc}} + V_{\text{in}}\left(\frac{t_1}{L_1}\right) \tag{3-82}$$

where I_{dc} is the dc current which exists as a consequence of setting the time period t_2 short of the value needed to just fully charge capacitor C_1.

Thus the time period t_1 is set to be slightly longer that the half cycle energy discharge time t_3, given by

$$t_3 = \pi\sqrt{L_{pe}C_1} \tag{3-83}$$

where L_{pe} is the primary leakage inductance given by

$$L_{pe} = L_p(1 - k^2) \tag{3-84}$$

The term I_m in Eq. 3-81 is the maximum current which would flow in the series circuit made of V_{in}, L_1, and C_1 and the primary of the transformer when the circuit is connected with zero initial current in L_1 and zero initial voltage on C_1, that is,

$$I_m = \frac{V_{in}}{Z} \tag{3-85}$$

and

$$Z = \sqrt{\frac{L_1}{C_1}} \tag{3-86}$$

for L_{pe}/L_1 is much less than 1.

3-11-2   Design Procedure for the Ward Converter

In order to design a particular converter, one specifies the desired output power P_{out}, defined ideally by

$$P_{out} = \frac{\frac{1}{2}C_1V_1^2f}{\eta} \tag{3-87}$$

where η is the converter efficiency and f is the operating frequency.

Select C_1 and V_1, choose the operating frequency (i.e., define the switching period T) and the t_3, which initially is assumed at one-half of T. A first approximation of the transformer parameters may be obtained using Eq. 3-80 and referring to Chap. 5.

Typically, V_1 will be in the range of two to six times V_{in}, so adjustments in the design may be necessary to bring it to the desired level of operation.

A critical factor in the successful operation of the Ward converter is the placement and precise selection of the capacitor C_2 across the output of the transformer. C_2 is given by the equation

$$C_2 = \frac{\mu C_1}{N^2} \tag{3-88}$$

where $0.5 \leq \mu \leq 1$ but closer to 0.5.

Another parameter depending on μ is defined as v and is given by

$$v = \frac{\mu}{1 + \mu} \tag{3-89}$$

which is required for defining the converter cycle period T. This expression then determines a value for L_1 and helps complete the specification of the converter.

The half-cycle period t_4 is given by

$$t_4 = \pi\sqrt{L_{pe}C_1}v$$

and

$$t_4 = t_3\sqrt{v} \tag{3-90}$$

which is chosen to be slightly greater than one-half t_3, which is the correct choice as can be seen by inspection of the waveforms of Fig. 3-26. For example, we can take $t_4 = 0.6t_3$, $v = 0.36$, and $\mu = 0.56$.

3-11-3 Features of the Ward Converter

The Ward converter was developed to satisfy a certain application need, which conventional topologies could not satisfy. The converter was used in a boost configuration, and the theoretically predicted waveforms presented here were reproduced by actual practical measurements.

Inherently, the converter has the feature of sustaining a shorted output without additional protection, making the converter useful in situations where capacitive loads are present. The converter also does its switching at zero current and therefore produces very little EMI and requires no snubber networks. The practical operating efficiency of this converter is also high, in the order of 85 to 90 percent. Another desirable feature of the converter is its ability to operate over a wide input voltage range.

As mentioned, this converter topology has not been fully analyzed; therefore, its potential usefulness has not been fully uncovered. Several practical aspects of the converter require further investigation, e.g., control-to-output transfer function, dynamic behavior, device stresses, and start-up transient. The inventor and this author believe that this is an intriguing topology, which brings new ideas and opens new horizons for further research by individuals who are dedicated to the power-conversion field. Such efforts are encouraged, and the reader may wish to obtain a copy of the patent or contact the inventor for a more detailed description. This author wishes to thank Dr. Michael Ward for his contribution.

REFERENCES

1. Bloom, G., and R. Severns: "Unusual DC-DC Power Conversion Systems," MIDCON, 1980.

2. Chryssis, G.: Power Supplies: A Switching Alternative, *Design News*, May 21, 1979.

3. Glen, D.: Self-Oscillating, Flyback Switching Converter, "Motorola TMOS Power FET Design Ideas," Handbook, 1985.

4. Hnatek, E. R.: "Design of Solid State Power Supplies," 2d ed., Van Nostrand Reinhold, New York, 1981.

5. Jarl, R. B., and D. R. Kemp: Application Note AN-6743, RCA, 1978.

6. Jones, D. V.: "A New Resonant Converter Topology," HFPC Proceedings, April 1987.

7. Lee, F. C.: "Zero Voltage Switching Techniques in DC-DC Converter Circuits," HFPC Proceedings, April 1987.

8. Middlebrook, R. D., and S. Ćuk: "Advances in Switched Mode Power Converters," vols. 1 and 2. Teslaco, Pasadena, Calif., 1981.

9. Motorola, Inc.: "Linear/Switchmode Voltage Regulator Handbook," 3d ed., 1987.

10. Pressman, A. I.: "Switching and Linear Power Supply, Power Converter Design," Hayden, Rochelle Park, N.J., 1977.

11. Sheppard, D. I., and B. E. Taylor: "A New Converter Topology Imparts Non-Pulsating Currents to Input and Output Power Lines," Powerconversion International, October 1984.

12. Suva, R., and R. J. Haver: "Engineering Bulletin EB-85, EB-86, EB-87, EB-88," Motorola, 1979.

13. Unitrode Corp.: "Linear Integrated Circuits Databook," 1987.

14. Velthooven, C. Van.: "Properties of DC-to-DC Converters for Switched Mode Power Supplies," Amperex Electronic Corp., 1975.

15. Ward, M.: "DC-to-DC Converter Current Pump," U.S. Patent Application No. 779790, September 1985.

16. Wong, T.: Compact Power Unit Raises Supply's Switching Rate by an Order of Magnitude, *Electronic Design*, April 1985.

17. Wood, P. N.: "Design Considerations for Transistor Converters," TRW Power Semiconductors, 1977.

18. ———: "Switching Power Converters," Van Nostrand Reinhold, New York, 1981.

FOUR

THE POWER TRANSISTOR
IN CONVERTER DESIGN

4-0 INTRODUCTION

In the block diagram of Fig. 1-1, depicting an off-the-line switching power supply, we show the block where conversion takes place containing what we called the switching element. Various types of switching elements, such as the transistor, the SCR, and the GTO, have been used by power supply designers over the years. By far the most popular and most often used element has been the bipolar transistor and in recent years its MOSFET counterpart. In this chapter we will discuss both types of transistors and the GTO, their characteristics, and their use in switch-mode power supplies.

4-1 TRANSISTOR SELECTION

The basic design parameters of a transistor to be used in an off-the-line converter are first its voltage blocking capability at turn-off and second its current carrying capacity at turn-on. Both these parameters are determined by the type of converter in which the transistor will be used. In Chap. 3 we described design equations and criteria for selecting the appropriate devices.

Another important decision which the designer has to face is whether to use bipolar transistors or MOSFETs in the design. Each of these transistors offers distinct advantages over the other, the bipolar being less expensive at present, while the MOSFET offers circuit simplicity because of simpler drive circuits.

It is important to note also that the bipolar transistor has a limited working frequency cutoff, which is around 50 kHz, while the MOSFET may be used in switching frequencies up to 200 kHz. Of course higher frequencies mean smaller component size and therefore more compact power supplies, a fact which seems to be the trend in today's power supply designs.

4-2 THE BIPOLAR POWER TRANSISTOR USED AS A SWITCH

The bipolar transistor is essentially a current-driven device; that is, by injecting a current into the base terminal a flow of current is produced in the collector. The amount of collector current flow is dependent upon the gain (beta) of the transistor, and the following relation holds true:

$$\beta = \frac{I_C}{I_B} \tag{4-1}$$

where I_C is the collector current (in amperes) and I_B is the base current (in amperes).

There are essentially two modes of operation in a bipolar transistor: the linear and saturating modes. The linear mode is used when amplification is needed, while the saturating mode is used to switch the transistor either on or off.

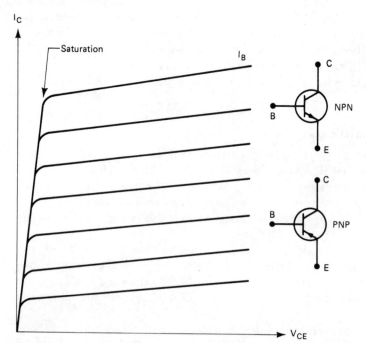

FIGURE 4-1 **Typical output characteristics and symbols of the bipolar transistor.**

Figure 4-1 shows the *V-I* characteristic of a typical bipolar transistor. Close examination of these curves shows that the saturation region of the *V-I* curve is of interest when the transistor is used in a switching mode. At that region a certain base current can switch the transistor on, allowing a large amount of collector current to flow, while the collector-to-emitter voltage remains relatively small.

In actual switching application a base drive current is needed to turn the transistor on, while a base current of reverse polarity is needed to switch the transistor back off. Since the transistor is a real-world element, it is far from ideal, and certain delays and storage times are associated with its operation.

In the following section are some definitions for a discrete bipolar transistor driven by a step function into a resistive load.

4-3 SWITCH TIMES DEFINITIONS OF BIPOLAR TRANSISTORS (RESISTIVE LOAD)

Figure 4-2 illustrates the base-to-emitter and collector-to-emitter wave forms of a bipolar NPN transistor driven into a resistive load by a base current

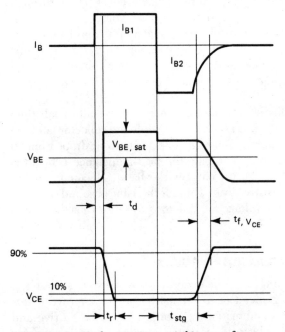

FIGURE 4-2   Bipolar transistor switching waveforms.

pulse I_B. The following are the definitions associated with these waveforms:

Delay Time, t_d   Delay time is defined as the interval of time from the application of the base drive current I_{B1} to the point at which the collector-emitter voltage V_{CE} has dropped to 90 percent of its initial off value.

Rise Time, t_r   Rise time is defined as the interval of time it takes the collector-emitter voltage V_{CE} to drop to 10 percent of its 90 percent off value.

Storage Time, t_{stg}   Storage time is the interval of time from the moment reverse base drive I_{B2} is applied to the point where the collector-emitter voltage V_{CE} has reached 10 percent of its final off value.

Fall Time, $t_{f,V_{CE}}$ Fall time is the time interval required for the collector-emitter voltage to increase from 10 to 90 percent of its off value.

4-4   INDUCTIVE LOAD SWITCHING RELATIONSHIPS

In the previous section, the definitions for the switching times of the bipolar transistor were made in terms of collector-emitter voltage. Since the load was defined to be a resistive one, the same definitions hold true for the collector current. However, when the transistor drives an inductive load, the collector voltage and current waveforms will differ. Since current through an inductor does not flow instantaneously with applied voltage, during turn-off, one expects to see the collector-emitter voltage of a transistor rise to the supply voltage before the current begins to fall. Thus, two fall time components may be defined, one for the collector-emitter voltage $t_{f,V_{CE}}$, and the other for the collector current t_{f,I_C}. Figure 4-3 shows the actual waveforms.

Observing the waveforms we can define the collector-emitter fall time $t_{f,V_{CE}}$ in the same manner as in the resistive case, while the collector fall time t_{f,I_C} may be defined as the interval in which collector current drops from 90 to 10 percent of its initial value. Normally the load inductance L behaves as a current source, and therefore it charges the base-collector transition capacitance faster than the resistive load. Thus for the same base and collector currents the collector-emitter voltage fall time $t_{f,V_{CE}}$ is shorter for the inductive circuit.

4-5   TRANSISTOR ANTISATURATION CIRCUITS

Figure 4-2 shows that the longest of the delay times is the storage time t_{stg}, and therefore the switching speed of the transistor would improve if this delay could be reduced. A combination of a large reverse base drive and antisaturation techniques may be used to reduce storage time to almost zero.

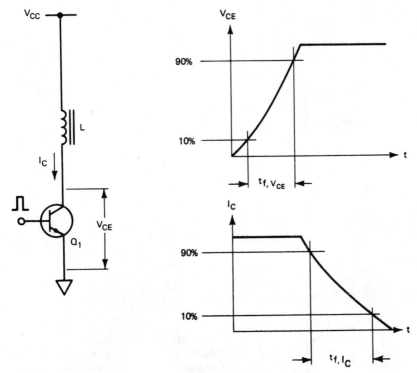

FIGURE 4-3 A bipolar switching transistor driving an inductive load, with the associated fall time characteristic waveforms. Note the current lagging the collector-emitter voltage.

The generation of the reverse base current using certain base drive techniques is discussed in Sec. 4-6. Here we will discuss two techniques used to keep the switching transistor out of saturation and ultimately improve their switching speed by reducing storage time to almost zero.

Figure 4-4a shows the use of antisaturation diodes, often called Baker clamps, in conjunction with a switching transistor. By inspection of the circuit, we observe that when the transistor is on, its base is two diode drops below the input. Assuming that diodes D_2 and D_3 have a forward-bias voltage of about 0.8 V, then the base will be 1.6 V below the input terminal. But because of diode D_1 the collector is one diode drop, or 0.8 V, below the input. Therefore, the transistor-collector will always be more positive than the base by $1.6 - 0.8 = 0.8$ V, hence out of saturation. Since the transistor normally works at high frequencies, 20 kHz and above, the antisaturation diodes must be fast recovery types. Diodes D_2 and D_3 may have low reverse blocking voltage ratings, but diode D_1 must have a rating of at least $2V_{CE}$. For an off-the-line power supply, a rating of 800 V PIV is normally used.

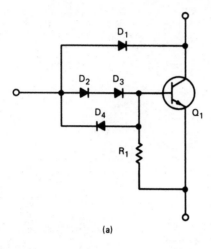

(a)

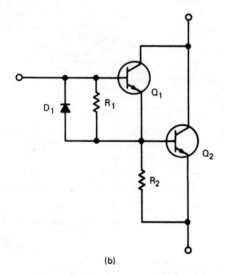

(b)

FIGURE 4-4 (a) Antisaturation diodes (Baker clamps) preventing transistor Q_1 from saturating; (b) the use of a Darlington circuit to keep Q_2 out of saturation.

Diode D_4 is a "wrap-around" type, which is used to pull the reverse base current at transistor turn-off, discharging the base-emitter capacitance and hence decreasing storage time.

Figure 4-4b shows a Darlington connection, which works in basically the same manner as described previously, with transistor Q_1 preventing Q_2 from fully saturating. An important point here is that Q_1 must turn off before Q_2 can begin to turn off. Diode D_1 provides a low impedance path for the reverse base current of Q_2 during turn-off. Resistors R_1 and R_2 are low-ohm resistors and provide a leakage current path for Q_1 and Q_2.

Darlington circuits may be implemented using discrete components, although monolithic Darlingtons are available from a number of power transistor manufacturers.

4-6 BASE DRIVE CIRCUIT TECHNIQUES FOR BIPOLAR TRANSISTORS

4-6-1 Constant Drive Current Circuits

In the previous paragraphs we examined the turn-on and turn-off limitations of the bipolar transistor when it is used as a power switch. It is now clear that in order to minimize the saturation losses, adequate forward base drive I_{B1} is required, while to minimize storage time and transistor switching losses, sufficient reverse base drive I_{B2} is needed.

We should note here that while I_{B2} is increased and the storage and fall times decrease, the emitter-to-base reverse bias voltage V_{EB} is also increased. This means that the reverse-bias secondary breakdown energy, known as E_{SB}, is also decreased, and if care is not taken in designing the reverse drive circuit, the switching transistor may be destroyed by going into a secondary breakdown mode. In Sec. 4-7-2 we will discuss the significance of E_{SB} and the phenomenon of secondary breakdown in bipolar transistors. Here it is sufficient to say that for practical purposes the reverse base drive circuit must have a low source impedance; that is, it must provide high I_{B2} and low V_{EB}.

The manufacturer's data sheets normally provide information on the limits of the reverse emitter-to-base bias voltage. Practical design circuits use a V_{EB} from -2 to -5 V. Higher reverse base voltage means a reduction in storage time delay because, by allowing fewer carriers to be neutralized by recombination, it takes a shorter period of time to remove the stored charge.

A popular base drive circuit used with a floating switching transistor is shown in Fig. 4-5 along with its waveforms. Operation of the circuit is as follows. When a positive secondary pulse V_S appears across the transformer, forward base current drive I_{B1} flows into the base of Q_1, turning the transistor on. Resistor R_1 limits this current within a predetermined value. This value of base current is determined using a forced gain ratio, which in practical circuits is between 8 and 10. Since the collector current may be easily calculated from the output power and type of converter used, the base current may therefore be predetermined using Eq. 4-1.

This positive drive pulse also charges capacitor C very rapidly. The charging voltage across the capacitor is

$$V_C = V_S - V_{BE} - V_D \tag{4-2}$$

where V_S = amplitude of transformer secondary voltage
V_{BE} = saturating base-emitter voltage of Q_1
V_D = forward-bias voltage of diode D

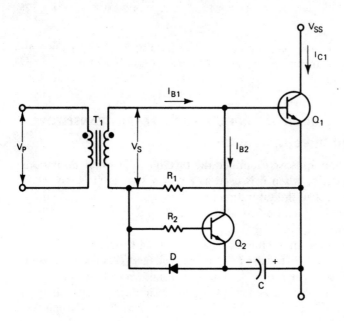

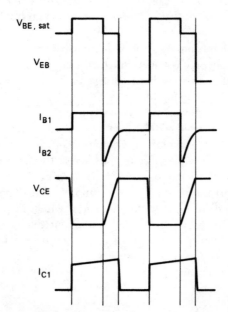

FIGURE 4-5 A base drive circuit using an isolation transformer driving the power switch Q_1 on and off. Typical switching waveforms of voltages and currents associated with the power switch are also shown.

If we assume $V_{BE} = V_D = 0.8$ V, then Eq. 4-2 becomes

$$V_C = V_S - 1.6 \qquad (4\text{-}3)$$

When the transformer primary goes to zero, the transformer secondary also goes to zero. Now capacitor C, which is fully charged, will forward-bias the base of Q_2, turning it on and consequently pulling the base of Q_1 to the negative potential.

With the capacitor now connected across the base-emitter junction of Q_1, a large reverse base current I_{B2} will develop. The magnitude of this current is determined by the capacitor and circuit resistances and the characteristics of transistors Q_1 and Q_2.

Another base drive circuit technique which has proved very effective in off-the-line switching power supplies is shown in Fig. 4-6. This circuit has the distinct advantage of providing adequate I_{B2} turn-off drive while using a minimum number of components. Circuit operation is as follows. When transistor Q_1 is turned on, the base drive transformer primary is connected to the supply voltage V_{CC}, storing energy in the transformer and also inducing a primary voltage pulse V_{P1}. This voltage pulse is transformer-coupled into the secondary. Since the transformer winding polarities are the same, a positive secondary voltage pulse V_S is generated, turning transistor Q_2 on.

The resistor R_1 is a base current limit resistor, designed to allow enough I_{B1} drive to force Q_2 on, without overdriving and oversaturating it. Resistor R_2 provides a leakage current path for the base-emitter junction of the switch, and it normally has a low ohm value, between 50 and 100 Ω.

When transistor Q_1 is switched off, the energy stored in the transformer is returned to V_{CC} by the tertiary winding and diode D_1. Since the polarity of this winding is opposite to the primary winding, a voltage pulse V_{P2} of reverse polarity is induced, coupling that negative pulse into the secondary and thus producing the desired reverse drive I_{B2} current flow.

When designing the base drive transformer, the primary-to-secondary turns ratio must be chosen not to exceed the transistor Q_2 published specifications for V_{BE} and V_{EB}. Normally the primary and tertiary winding turns are the same.

It should also be noted that the primary and tertiary windings must be tightly wound (i.e., bifilar) to avoid excess voltage spikes due to leakage inductance. Transistor Q_1 must be chosen to withstand a collector turn-off voltage of at least $2V_{CC}$. A modification of the above circuit will make the transformer even simpler, while retaining the advantages described previously. A practical implementation is shown in Fig. 4-7.

If a positive pulse V_P is applied to the primary winding of the base drive transformer, a secondary voltage pulse V_S is induced, turning transistor Q_1 on. During the duration of the on pulse the forward drive I_{B1} charges capacitor C with the polarity shown in Fig. 4-7. The voltage across the capacitor

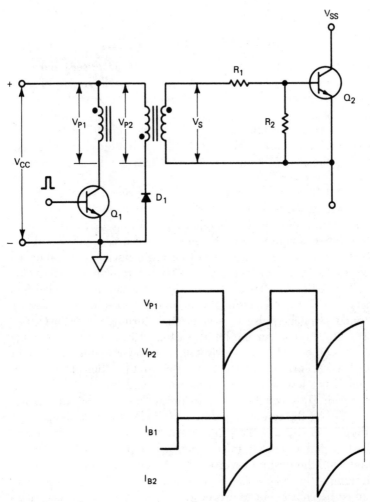

FIGURE 4-6 A transformer-coupled base drive circuit, which uses the energy stored in the transformer to generate the reverse base drive through the tertiary winding.

is clamped to about 3 V, determined by the forward voltage drops of diodes D_1, D_2, and D_3. Instead of diodes a zener of appropriate voltage rating could be used. When the primary voltage goes to zero, the transformer secondary voltage will also be zero. At this time the positive terminal of capacitor C is at the transistor Q_1 emitter potential. Thus the charged capacitor is now effectively connected across the base-emitter junction of the switch, producing the necessary reverse base drive I_{B2} to turn the transistor off and to reduce its storage time.

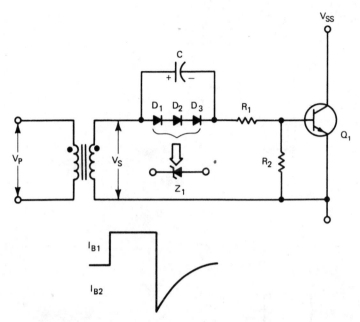

FIGURE 4-7 A base drive circuit using a simple isolation transformer to generate the transistor on pulse. Turn-off drive is generated by the negative charge of capacitor C.

Because of the simplicity of this scheme it may also be used to drive direct-coupled power transistors, as shown in Fig. 4-8. An emitter-follower comprised of transistors Q_1 and Q_2 alternately switches the base of Q_3 to V_{CC} or ground, thus turning it on or off. When Q_1 is on, transistor switch Q_3 turns on, also charging capacitor C as shown. Zener diode Z_1 limits the charge on C (in practical circuits $V_Z = 4.3$ V) and also provides a path for the forward base drive current I_{B1}, which is set by resistor R_1. With transistor Q_1 off and Q_2 on, the charged capacitor C is effectively connected to the base-emitter of Q_3, causing reverse current I_{B2} to flow because of the polarity of the capacitor. The value of I_{B2} depends on the gain of Q_2, the value and therefore the charge on capacitor C, and the circuit impedances.

4-6-2　A Proportional Base Drive Circuit

All the base drive circuits described in the previous section provide a constant drive current to the transistor switch. These circuits have the drawback that at low collector current, the transistor storage time may not be shortened sufficiently and effectively because of a change in transistor β.

Using the proportional base drive technique we can control the value of β and in fact we can keep it constant for all collector currents. Therefore,

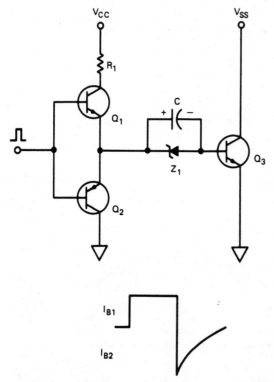

FIGURE 4-8 A capacitor-coupled direct drive base circuit. Forward and reverse base drive currents are generated by the charged capacitor C. Zener Z_1 is used to clamp the voltage across the capacitor to a predetermined level.

in this type of drive, one expects shorter storage times, at low collector currents, than those achieved by the constant drive current techniques.

Figure 4-9 shows a possible way of realizing a proportional base drive circuit. The circuit operates as follows. When transistor Q_1 turns on, transformer T_1 is in negative saturation and transistor Q_2 is off. In the time interval where Q_1 is on, a current flows through winding N_1, limited by series resistor R, storing energy in the winding and also keeping T_1 in saturation. When transistor Q_1 turns off, the energy stored in N_1 is transferred to winding N_4, causing base current to flow in Q_2, turning it on. With transistor Q_2 on, collector current I_C flows, energizing transformer winding N_2. Therefore all the dotted ends of transformer T_1 become positive, pulling the core from negative to positive saturation.

Since windings N_2 and N_4 are now acting as a current transformer, transistor Q_2 stays on, forcing a constant β at all collector current levels. Q_2 turns

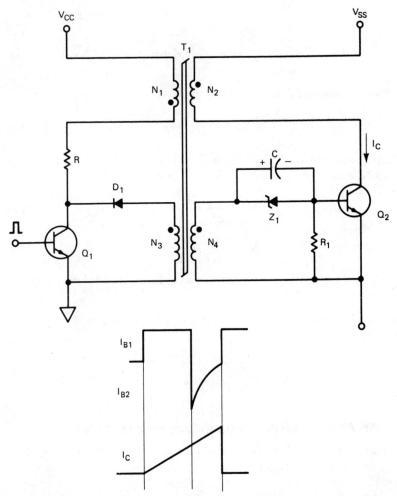

FIGURE 4-9 A proportional base drive circuit with base current and collector current waveforms shown.

off by turning Q_1 on. The following equations may be used to calculate the transformer turns ratio. Using a forced constant β value for Q_2 we have

$$\beta = \frac{N_4}{N_2} \qquad (4\text{-}4)$$

During transformer operation the flux density at t_{on} must be equal to the flux density at t_{off}.

$$\Delta \Phi t_{on} = \Delta \Phi t_{off} \qquad (4\text{-}5)$$

and

$$\Delta\Phi = 2B_{max}A_C \qquad (4\text{-}6)$$

where B_{max} is the maximum operating flux density in gauss, and A_C is the core area (in square centimeters).

From the fundamental magnetic equation we get

$$N(\Delta\Phi) = \frac{V}{2f(10^{-8})} \qquad (4\text{-}7)$$

Combining and equating Eqs. 4-6 and 4-7 we can write for transformer windings N_1 and N_4 at 50% duty cycle maximum

$$N_1 = \frac{V_{CC}(10^8)}{2fB_{max}A_C} \qquad (4\text{-}8)$$

$$N_4 = \frac{V_{BC}(10^8)}{2fB_{max}A_C} \qquad (4\text{-}9)$$

where V_{BE} is the base emitter voltage of transistor Q_2, and f is the converter operating frequency (in kilohertz).

An expression of the turns ratio N_1/N_4 may also be derived by dividing Eqs. 4-8 and 4-9.

$$\frac{N_1}{N_4} = \frac{V_{CC}}{V_{BE}} \frac{t_{off}}{t_{on}} \qquad (4\text{-}10)$$

4-6-3 An Alternative Proportional Base Drive Circuit

The circuit depicted in Fig. 4-10 is an alternative proportional base drive circuit. The author has used this circuit in real applications and has found it to perform very well, when carefully designed.

Operation of the circuit is as follows: A positive control pulse at the base of transistor Q_1 turns it on. Because of the polarity of the base drive transformer windings, transistor Q_2 is off. At the same time magnetizing current I_d is building up in N_d, approaching a steady state. The final value of this current will be

$$I_d = \frac{V_{dd}}{R} \qquad (4\text{-}11)$$

When Q_1 turns off, I_d ceases, and because of the T_1 transformer windings polarity, energy is transferred to Q_2 base winding N_b, inducing a current I_b to flow, turning Q_2 on.

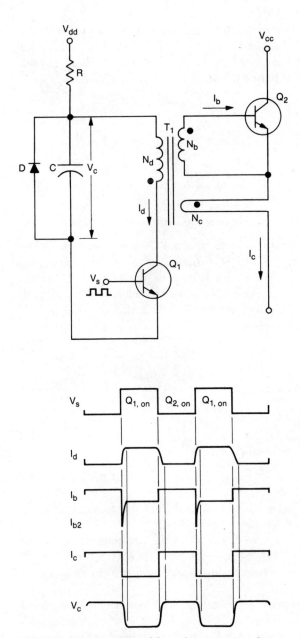

FIGURE 4-10 Proportional base drive circuit and its associated waveforms.

Hence, current I_c is now flowing through winding N_c. This current flow causes a regenerative increase in base drive to Q_2, since windings N_b and N_c act as a current transformer, until Q_2 is switched fully on. The final value of I_c induces a proportional base current given by the equation

$$I_c = I_b \frac{N_b}{N_c} \tag{4-12}$$

and

$$\frac{N_d}{N_b} = \frac{V_{dd} - 1}{V_{bb}} \tag{4-13}$$

where V_{bb} is the Q_2 turn-off base drive source voltage at maximum I_c.

While transistor Q_2 is on and Q_1 is off, capacitor C charges through resistor R to the supply voltage V_{dd}. When Q_1 turns on again, capacitor C applies its charge to winding N_d, driving the voltage at the base of Q_2 sharply negative, thus providing the reverse base drive turn-off current I_{b2}, rapidly turning off transistor Q_2. The energy which the capacitor C has to provide to turn off Q_2 is given by the equation

$$W = \tfrac{1}{2}C(V_{dd} - 1)^2 \tag{4-14}$$

and

$$C \approx \frac{2(I_d)t_f}{V_{dd} - 1} \tag{4-15}$$

where t_f is the published transistor fall time.

Any remaining voltage on capacitor C helps to rebuild the magnetizing current I_d, thus repeating the cycle.

Diode D is used to clamp any underdamped ringing that could cause the upper end of winding N_d to become negative.

Figure 4-11 depicts an improved version of the previous circuit, which makes its use practical at high frequencies. In this circuit transistor, Q_3 and its associated components compromise a fast discharge circuit for capacitor C.

During the time period that Q_2 is on and Q_1 is off, current flowing through resistor R is multiplied by the gain β of transistor Q_3, which significantly reduces the charging time of capacitor C. Diode D_2 is used to discharge the capacitor C when transistor Q_1 turns on.

In this design it is desirable to set the operating point of the transformer near saturation, that is, $B_{max} \leq B_{sat}$.

Turns ratios for the drive transformer were established by Eqs. 4-12 and

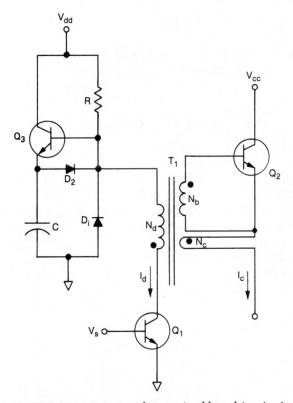

FIGURE 4-11 An improved proportional base drive circuit.

4-13. The drive transformer primary winding N_d is given by the equation

$$N_d = \frac{V_{dd}(10^8)}{2f(B_{max})A_c} \qquad (4\text{-}16)$$

where A_c is the chosen core area in square centimeters.

After the turns ratios of the transformer are calculated, find the magnetic path length of the desired core using the equation

$$l_i = \frac{H}{N_d I_d} \qquad (4\text{-}17)$$

where H is the coercive force corresponding to the chosen B_{max} value of the core to be used.

If the actual l_i of the selected core is smaller than the calculated magnetic path length, the core will be heavily saturated and will not store enough energy to provide the desired base drive currents. Either use a larger core, or introduce a small gap l_g using the equation

$$l_g = \frac{l_e - l_i}{\mu_a} \tag{4-18}$$

where $\mu_a = B/H$ the permeability of the core, l_i is the length of magnetic path of material, and l_e is the effective magnetic path length.

4-6-4 Antisaturation Circuits Used in Base Drives

In Sec. 4-5 we discussed two methods used to keep the switching transistor of a power converter out of saturation, thus reducing the storage time to negligible values. These antisaturation circuits may be incorporated into the base drive circuits presented in the preceding discussions with excellent results. Figure 4-12 shows a typical application using Baker clamps in a basic base drive circuit. All other base drive circuits described thus far may also be easily adapted to this scheme. Of course, if the switching transistor is a Darlington, no antisaturation diodes are needed, since the Darlington inherently presents antisaturation features.

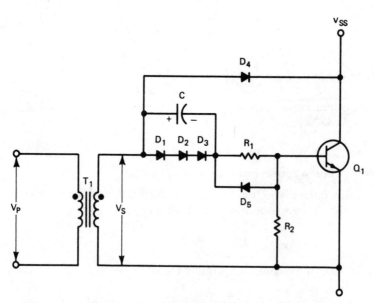

FIGURE 4-12 Base drive circuit shown in Fig. 4-6, redrawn here to show antisaturation diodes (Baker clamps), which reduce the storage time of transistor Q_1 by keeping it out of saturation.

4-7 BIPOLAR TRANSISTOR SECONDARY BREAKDOWN CONSIDERATIONS

4-7-1 Forward-Bias Secondary Breakdown

Thus far we have observed during our discussion of power converters that the switching transistor is subjected to great stress, during both turn-on and turn-off. It is imperative that the engineer clearly understands how the power bipolar transistor behaves during forward- and reverse-bias periods in order to design reliable and trouble-free circuits.

The first problem is to avoid secondary breakdown of the switching transistor at turn-on, when the transistor is forward-biased. Normally the manufacturer's specifications will provide a safe-operating area (SOA) curve, such as the typical one shown in Fig. 4-13. In this figure collector current is plotted against collector-emitter voltage. The curve locus represents the maximum limits at which the transistor may be operated. Load lines that fall within the pulsed forward-bias SOA curve during turn-on are considered safe, provided that the device thermal limitations and the SOA turn-on time are not exceeded.

The phenomenon of forward-biased secondary breakdown is caused by hot spots which are developed at random points over the working area of a power transistor, caused by unequal current conduction under high-voltage

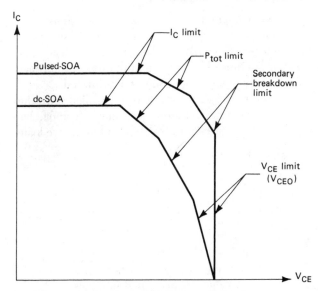

FIGURE 4-13   DC and pulsed SOA curves for bipolar power transistors.

stress. Since the temperature coefficient of the base-to-emitter junction is negative, hot spots increase local current flow. More current means more power generation, which in turn raises the temperature of the hot spot even more. Since the temperature coefficient of the collector-to-emitter breakdown voltage is also negative, the same rules apply. Thus if the voltage stress is not removed, ending the current flow, the collector-emitter junction breaks down and the transistor fails because of thermal runaway.

Recently a method for avoiding forward breakdown has been developed by National Semiconductor using a modified emitter-ballasting technique during transistor manufacturing. Devices manufactured using this technique may be operated at maximum rated power levels and collector voltages with no fear of secondary breakdown. Figure 4-14 shows a schematic of the complete monolithic device.

This technique implants a junction field-effect transistor (JFET) in series with the power transistor base. The JFET acts as a base ballast resistor whose resistance changes as a function of collector-to-base voltage. This method differs from the standard emitter ballasting, where the resistor is placed in series with the device emitter. Base ballasting also maintains constant power dissipation regardless of collector voltage. Resistor R (Fig. 4-14) takes over as the JFET pinches off.

4-7-2 Reverse-Bias Secondary Breakdown

It was mentioned in previous paragraphs that when a power transistor is used in switching applications, the storage time and switching losses are the two most important parameters with which the designer has to deal extensively. If storage time is not minimized, saturation of the transformer takes place, and also the range of regulation of the converter is limited.

On the other hand the switching losses must also be controlled since they affect the overall efficiency of the system. Figure 4-15 shows turn-off characteristics of a high-voltage power transistor in resistive and inductive loads.

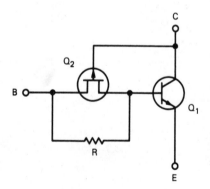

FIGURE 4-14 Secondary breakdown is prevented in bipolar power transistors by integrating a JFET in series with the base. The JFET acts as a ballast resistor.

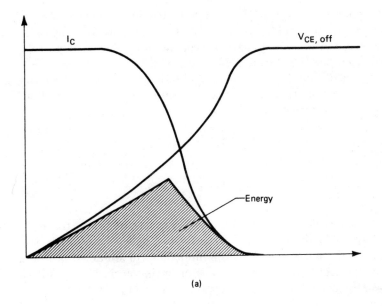

(a)

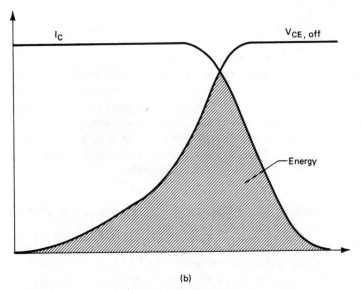

(b)

FIGURE 4-15 Turn-off characteristics of a high-voltage power transistor for a resistive (*a*) and an inductive (*b*) load. Cross-hatched area in each curve represents the switching loss energy.

Inspecting the curves we can see that the inductive load generates a much higher peak energy at turn-off than its resistive counterpart. It is then possible, under these conditions, to have a secondary breakdown failure if the reverse-bias safe operating area (RBSOA) is exceeded.

In early transistor literature the method of rating reverse-bias secondary breakdown was to test the transistor using an unclamped inductive load. The reverse-bias secondary breakdown energy E_{SB} was calculated as

$$E_{SB} = \tfrac{1}{2} L_{eff} I_C^2 \qquad (4\text{-}19)$$

where

$$L_{eff} = \frac{V_{CEX}}{V_{CEX} - V_{CC}} L \qquad (4\text{-}20)$$

The calculated E_{SB} is given in joules. But, since actual testing of the transistor may be performed with either open base turn-off or with very high base impedance, the E_{SB} range may vary from millijoules to joules. Also, taking into account the fact that the transistor is operating near the breakdown voltage V_{CEX}, the parameter of E_{SB} is relatively useless to modern transistor specifications.

An alternative RBSOA rating system has been developed by power transistor manufacturers using a clamped inductive collector load. The resulting curve is shown in Fig. 4-16, and in a way it resembles the forward-bias SOA curve. The RBSOA curve shows that for voltages below V_{CEO} the safe area is independent of reverse-bias voltage V_{EB} and is only limited by the device collector current I_C. Above V_{CEO} the collector current must be derated depending upon the applied reverse-bias voltage.

It is then apparent that the reverse-bias voltage V_{EB} is of great importance and its effect on RBSOA very interesting. It is also important to remember that avalanching the base-emitter junction at turn-off must be avoided, since turn-off switching times may be decreased under such conditions. In any case, avalanching the base-emitter junction may not be considered relevant, since normally designers protect the switching transistors with either clamp diodes or snubber networks to avoid such encounters.

4-8 SWITCHING TRANSISTOR PROTECTIVE NETWORKS: *RC* SNUBBERS

It is now very clear from the previous discussion that the most critical portion of the switching cycle occurs during transistor turn-off. During the presentation of the base drive techniques we mentioned that normally reverse-base current I_{B2} is made very large in order to minimize storage time. Unfortunately this condition may avalanche the base-emitter junction and destroy the transistor. Alternatively there are two options to prevent this from happening: (1) turning off the transistor at low values of collector-to-

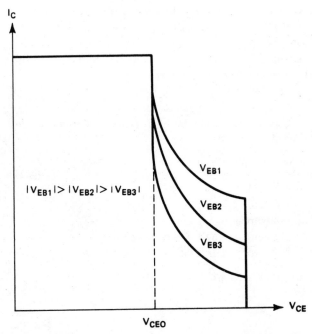

FIGURE 4-16 Reverse-bias safe operating area (RBSOA) plot for high-voltage switching transistors as a function of reverse-bias voltage V_{EB}.

emitter voltage V_{CE} and (2) reducing collector current with rising collector voltage.

Of course when the power supply design is an off-the-line type, the second solution seems to be the more realistic one. Figure 4-17 shows how this can be accomplished by using an RC snubber network across the transistor to divert collector current during turn-off. The circuit works as follows. When

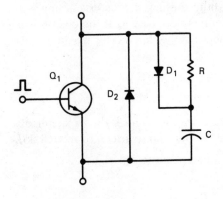

FIGURE 4-17 A turn-off current snubber network across a switching bipolar transistor. Diode D_2 is a leakage inductance commutating diode.

transistor Q_1 is off, capacitor C is charged through diode D_1 to a voltage $(V_{CC} - V_D)$. When Q_1 turns on, the capacitor discharges through resistor R. It is important to note that the snubber actually dissipates a fair amount of power, thus relieving the switching transistor, which would have to dissipate this power in the absence of the snubber.

The following analysis and design procedure is based on quasiempirical results, but the equations may be successfully used to develop snubber networks for practical designs. Referring to Fig. 4-15b, the energy area during turn-off may be written as

$$E = \frac{CV_{CE}^2}{2} = \frac{I_C V_{CE}(t_r + t_f)}{2} \tag{4-21}$$

where
I_C = maximum collector current, A
V_{CE} = maximum collector-emitter voltage, V
t_r = maximum collector voltage rise time, μs
t_f = maximum collector current fall time, μs

Solving Eq. 4-21 for capacitance C, we get

$$C = \frac{I_C(t_r + t_f)}{V_{CE}} \tag{4-22}$$

As stipulated earlier, capacitor C charges at turn-off and discharges through R during the transistor turn-on time t_{on}. The voltage across the capacitor may be written as

$$V_C = V_{CE} \exp - (t_{on}/RC) \tag{4-23}$$

In order to guarantee a fully charged capacitor prior to turn-off near V_{CE}, we must choose RC such that the expression $\exp - (t_{on}/RC)$ approaches unity. By the same token, we must also choose RC such that the capacitor will be discharged at the end of the turn-on time t_{on}.

From basic circuit theory we know that it takes five time constants (5τ; $\tau = RC$) for a capacitor to discharge fully through a resistor. Assuming in this case that the capacitor will be essentially discharged at the end of three time constants, the following expression may be derived for maximum discharge resistor value:

$$R = \frac{t_{on}}{3C} \tag{4-24}$$

With value calculated for R in Eq. 4-24, we must check the capacitor discharge current through the transistor at turn-on and restrict it to about $0.25I_C$ using the following formula:

$$I_{dis} = \frac{V_{CE}}{R} \tag{4-25}$$

If the resistor is too low and $I_{dis} > 0.25 I_C$, then R may arbitrarily be raised to fulfill the constraint.

The last step is to calculate the maximum resistor power rating given by

$$P_R = \tfrac{1}{2} C V_{CE}^2 f \qquad (4\text{-}26)$$

where f is the converter working frequency in kilohertz.

The following example gives a numerical verification of the above formulas.

EXAMPLE 4-1

Consider a switching transistor used in a half-bridge converter where $V_{CE} = 200$ V, $t_f = 2\ \mu s$, and $t_r = 0.5\ \mu s$. The converter is working at 20 kHz, and the transistor collector current is $I_C = 2$ A. Calculate the resistance R and the capacitance C of the snubber network.

SOLUTION

From Eq. 4-22 we have

$$C = \frac{I_C(t_r + t_f)}{V_{CE}} = \frac{2(0.5 + 2) \times 10^{-6}}{200} = 0.025\ \mu F = 25\ nF$$

We will use 22 nF. Assume that t_{on} is 40 percent of the total period $(1/f)$. Then

$$t_{on} = \frac{0.4 \times 10^{-3}}{20} = 0.02 \times 10^{-3} = 20\ \mu s$$

Using Eq. 4-24,

$$R = \frac{20 \times 10^{-6}}{3(0.22) \times 10^{-6}} = 303\ \Omega$$

We will use 300 Ω.

Check for the discharge current

$$I_{dis} = \frac{200}{300} = 0.67\ A$$

This is greater than 25 percent of I_C, thus a new R must be calculated.

$$R = \frac{V_{CE}}{0.25 I_C} = \frac{200}{(0.25)(2)} = 400\ \Omega$$

Take $R = 430\ \Omega$.

Finally the power rating of the resistor is calculated.

$$P_R = \frac{(0.025 \times 10^{-6})(200^2)(2 \times 10^3)}{2} = 1\ W$$

4-9 THE POWER MOSFET USED AS A SWITCH

4-9-1 Introduction

Although the field-effect transistor (FET) has been used in circuit designs for many years, the power metal-oxide-semiconductor field-effect transistor (MOSFET) has been perfected in recent years to make it commercially available for power electronics designs. The MOSFET was developed out of the need for a power device that could work beyond the 20-kHz frequency spectrum, anywhere from 100 kHz to above 1 MHz, without experiencing the limitations of the bipolar power transistor.

Of course there are several advantages to designing converters working at, say, 100 kHz rather than 20 kHz, the most important being reduced size and weight. The power MOSFET offers the designer a high-speed, high-power, high-voltage device with high gain, almost no storage time, no thermal runaway, and inhibited breakdown characteristics. Different manufacturers use different techniques for constructing a power FET, and names like HEXFET, VMOS, TMOS, etc., have become trademarks of specific companies. The bottom line is that all MOSFETs work on the same principle, although their variation in construction may enhance certain performance specifications, a fact which might make a specific type of MOSFET more attractive than others for some applications.

4-9-2 Basic MOSFET Definitions

The circuit symbol for a MOSFET is shown in Fig. 4-18. This is an N-channel MOSFET, and its counterpart NPN bipolar transistor is also depicted for comparison purposes. Of course there is also a P-channel MOSFET in which the arrowhead is pointing the opposite way. Looking at the

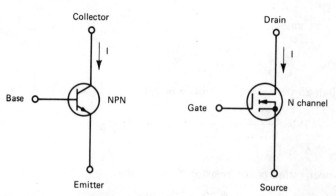

FIGURE 4-18 Electrical symbols of an NPN bipolar transistor and its equivalent N-channel MOSFET.

two symbols of Fig. 4-18, the collector, base, and emitter terminals of the bipolar transistor are termed drain, gate, and source, respectively, in the MOSFET.

Although both devices are called transistors, it is important to understand that there are distinct differences in the construction and principles of operation between bipolar and MOSFET devices. The first and most important difference is the fact that the MOSFET is a majority carrier semiconductor device, while the bipolar is a minority carrier semiconductor.

4-9-3 Gate Drive Considerations of the MOSFET

When the bipolar power transistor was examined, we mentioned the fact that this device is essentially current-driven, that is, a current must be injected at the base in order to produce a current flow in the collector. This current flow in turn is proportional to the gain of the bipolar transistor.

The MOSFET, on the other hand, is a voltage-controlled device; that is, a voltage of specified limits must be applied between gate and source in order to produce a current flow in the drain. Since the gate terminal of the MOSFET is electrically isolated from the source by a silicon oxide layer, only a small leakage current flows from the applied voltage source into the gate. Thus, we can say that the MOSFET has an extremely high gain and high impedance.

In order to turn a MOSFET on, a gate-to-source voltage pulse is needed to deliver sufficient current to charge the input capacitor in the desired time. The MOSFET input capacitance C_{iss} is the sum of the capacitors formed by the metal-oxide gate structure, from gate to drain (C_{GD}) and gate to source (C_{GS}). Thus, the driving voltage source impedance R_g must be very low in order to achieve high transistor speeds.

A way of estimating the approximate driving generator impedance, plus the required driving current, is given in the following equations:

$$R_g = \frac{t_r \text{ (or } t_f)}{2.2C_{iss}} \qquad (4\text{-}27)$$

and

$$I_g = C_{iss} \frac{dv}{dt} \qquad (4\text{-}28)$$

where R_g = generator impedance, Ω
  C_{iss} = MOSFET input capacitance, pF
  dv/dt = generator voltage rate of change, V/ns

To turn off the MOSFET, we need none of the elaborate reverse current generating circuits described for bipolar transistors. Since the MOSFET is a majority carrier semiconductor, it begins to turn off immediately upon removal of the gate-to-source voltage. Upon removal of the gate voltage the transistor shuts down, presenting a very high impedance between drain and source, thus inhibiting any current flow, except leakage currents (in microamperes). Figure 4-19 illustrates the relationship of drain current versus drain-to-source voltage. Note that drain current starts to flow only when the drain-to-source avalanche voltage is exceeded, while the gate-to-source voltage is kept at 0 V.

EXAMPLE 4-2

A power MOSFET driven from a 12-V dc generator is switching 320 V dc of supply voltage. Given that the MOSFET has a $C_{GD} = 100$ pf and $C_{GS} = 500$ pf, find the total gate current I_g required to switch the transistor on within 20 ns.

SOLUTION

From Eq. 4-28 we deduce that in order to charge the gate-to-drain capacitance C_{GD}, a current of magnitude $C_{GD}(dv/dt)$ is required to overcome the Miller effect. Therefore,

$$I_m = C_{GD} \frac{dv}{dt} = 100 \text{ pf} \left(\frac{320 \text{ V}}{20 \text{ ns}} \right) = 1.6 \text{ A}$$

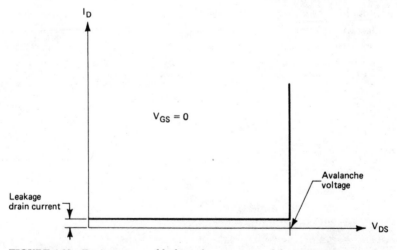

FIGURE 4-19 Drain-to-source blocking characteristics of the MOSFET. Note that when the avalanche voltage is reached, the drain current increases linearly.

Since by design the MOSFET responds almost instantaneously to gate voltage, it will conduct with a V_{GS} somewhere between 2 to 4 V, and it will be fully on with a gate voltage of 6 to 8 V. Assuming for this example that the MOSFET under consideration is fully on at 7 V, the current I_C required to charge the gate-to-source capacitance C_{GS} is

$$I_C = C_{GS} \frac{dv}{dt} = 500 \text{ pf} \left(\frac{7 \text{ V}}{20 \text{ ns}} \right) = 0.175 \text{ A}$$

Therefore, the total generator current I_g required to turn the FET on is

$$I_g = I_m + I_c = 1.775 \text{ A}$$

4-9-4 Static Operating Characteristics of the MOSFET

Figure 4-20 shows the drain-to-source operating characteristics of the power MOSFET. The reader may compare these curves to the ones given in Fig. 4-1 for the bipolar transistor. Although the two curves look the same, there are some fundamental differences between them.

The MOSFET output characteristic curves reveal two distinct operating regions, namely, a "constant resistance" and a "constant current." Thus as

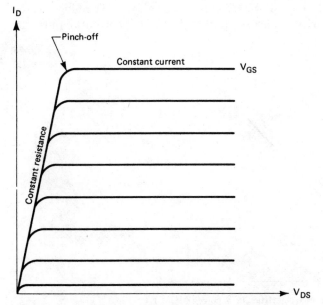

FIGURE 4-20 Typical output characteristic curves of a MOSFET.

the drain-to-source voltage is increased, the drain current increases proportionally, until a certain drain-to-source voltage called "pinchoff" is reached. After pinchoff, an increase in drain-to-source voltage produces a constant drain current.

When the power MOSFET is used as a switch, the voltage drop between the drain and source terminals is proportional to the drain current; that is, the power MOSFET is working in the constant resistance region, and therefore it behaves essentially as a resistive element. Consequently the on-resistance $R_{DS,\text{on}}$ of the power MOSFET is an important figure of merit because it determines the power loss for a given drain current, just as $V_{CE,\text{sat}}$ is of importance for the bipolar power transistor. By examining Fig. 4-20, we note that the drain current does not increase appreciably when a gate-to-source voltage is applied; in fact, drain current starts to flow after a threshold gate voltage has been applied, in practice somewhere between 2 and 4 V. Beyond the threshold voltage, the relationship between drain current and gate voltage is approximately equal. Thus, the transconductance g_{fs}, which is defined as the rate of change of drain current to gate voltage, is practically constant at higher values of drain current. Figure 4-21 illustrates the transfer characteristics of I_D vs. V_{GS}, while Fig. 4-22 shows the relationship of transconductance g_{fs} to drain current.

It is now apparent that a rise in transconductance results in a proportional rise in the transistor gain, i.e., larger drain current flow, but unfortunately this condition swells the MOSFET input capacitance. Therefore, carefully designed gate drivers must be used to deliver the current required to

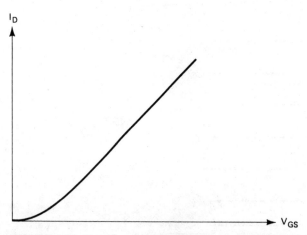

FIGURE 4-21 Transfer characteristics of a power MOSFET showing the linear $d\mathbf{I}_D/d\mathbf{V}_{GS}$ relationship.

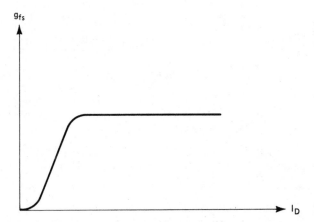

FIGURE 4-22 Curve showing the relationship of transconductance g_{fs} to drain current. Note how rapidly the transconductance rises to saturation as drain current is increased.

charge the input capacitance in order to enhance the switching speed of the MOSFET.

4-9-5 MOSFET Safe Operating Area (SOA)

In the discussion of the bipolar power transistor, it was mentioned that in order to avoid secondary breakdown, the power dissipation of the device must be kept within the operating limits specified by the forward-bias SOA curve. Thus at high collector voltages the power dissipation of the bipolar transistor is limited by its secondary breakdown to a very small percentage of full rated power. Even at very short switching periods the SOA capability is still restricted, and the use of snubber networks is incorporated to relieve transistor switching stress and avoid secondary breakdown.

In contrast, the MOSFET offers an exceptionally stable SOA, since it does not suffer from the effects of secondary breakdown during forward bias. Thus both the dc and pulsed SOA are superior to that of the bipolar transistor. In fact with a power MOSFET it is quite possible to switch rated current at rated voltage without the need of snubber networks. Of course, during the design of practical circuits, it is advisable that certain derating must be observed. Figure 4-23 shows typical MOSFET and equivalent bipolar transistor curves superimposed in order to compare their SOA capabilities.

Secondary breakdown during reverse bias is also nonexistent in the power MOSFET, since the harsh reverse-bias schemes used during bipolar transistor turn-off are not applicable to MOSFETs. Here, for the MOSFET to turn off, the only requirement is that the gate is returned to 0 V.

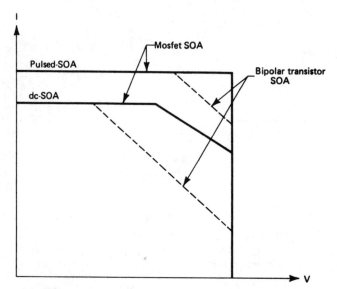

FIGURE 4-23 DC and pulsed SOA curves for a MOSFET power transistor (solid lines). SOA curves of an equivalent bipolar transistor are also shown (dotted lines). Note the superior SOA characteristics of the MOSFET device.

4-9-6 Design Considerations for Driving the Power MOSFET

It must be clear by now that using the power MOSFET the designer may achieve performance much superior to using bipolar power transistors. Since the best performance characteristics of the MOSFET come forth when the device is operated at very high frequencies (normally 100 kHz and above), certain design precautions must be taken in order to minimize problems, especially oscillations. Figure 4-24 shows a typical MOSFET driving a resistive load, working in the common-source mode.

There are basically two very simple design rules associated with MOSFET application which will prevent the transistor from oscillating when used in high frequencies. First, minimize all lead lengths going to the MOSFET terminals, especially the gate lead. If short leads are not possible, then the designer may use a ferrite bead or a small resistor R_1 in series with the MOSFET as shown in Fig. 4-24. Either one of those elements when placed close to the transistor gate will suppress parasitic oscillations.

Second, because of the extremely high input impedance of the MOSFET, the driving source impedance must be low in order to avoid positive feedback which may lead to oscillations. We must note at this point also that while the dc input impedance of the MOSFET is very high, its dynamic or ac input impedance varies with frequency. Therefore, the rise and fall times of the MOSFET depend on the driving generator impedance.

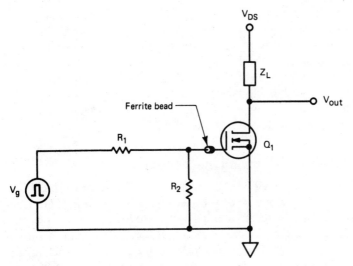

FIGURE 4-24 A typical MOSFET used as a switch, operating in common-source configuration.

An approximation of the rise and fall times is given by the following equation:

$$t_r \text{ or } t_f = 2.2 R_g C_{iss} \qquad (4\text{-}29)$$

where t_r = MOSFET rise time, ns
   t_f = MOSFET fall time, ns
   R_g = driving generator impedance, Ω
   C_{iss} = MOSFET input capacitance, pF

It is important to note that Eq. 4-29 is valid if $R_L \gg R_g$. This information, along with the fact that there are no storage or delay times associated with the MOSFET, allows the rise and fall times to be set by the designer. The resistor R_2 in the circuit of Fig. 4-24 is used to assist transistor turn-off.

EXAMPLE 4-3

In the circuit of Fig. 4-24, a MOSFET is used which has an input capacitance C_{iss} = 500 pF, resistor R_1 = 150 Ω, and R_L = 2000 Ω. What will be the rise time of the driving waveform?

SOLUTION

Using Eq. 4-29 we have

$$t_r = (2.2)(150)(500 \times 10^{-12}) = 165 \text{ ns}$$

Another important thing to remember is the fact that the silicon oxide layer between the gate and source regions can be easily perforated and therefore permanently destroyed if the gate-to-source voltage exceeds manufacturer's specifications. Practical gate voltages have a maximum value anywhere from 20 to 30 V. Even if the gate voltage is below the maximum permissible value, it is advisable to perform a thorough investigation to make sure that there are not any fast rising spikes, caused by stray inductances, which may destroy the oxide layer of the MOSFET.

4-9-7 Circuits Used in Driving the MOSFET

Driving the MOSFET from TTL   Although it is possible to drive the MOSFET directly from the output of some transistor-transistor logic (TTL) families, direct driving is not recommended, since the transistor stays in the linear region for a long time before reaching saturation. Thus the performance of the MOSFET may never reach its optimum point with such a gate drive.

In order to improve the switching performance, a buffer circuit must be provided, which will present very fast current sourcing and sinking to the gate capacitances. Such a simple circuit is a complementary emitter-follower stage, as shown in Fig. 4-25. Transistors Q_1 and Q_2 must be chosen to have

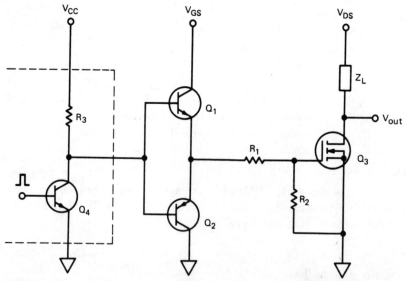

FIGURE 4-25   An emitter-follower buffer is used between TTL and MOSFET to decrease switching rise and fall times. These transistors must have high gain and wide bandwidth.

high gain at high current in order to be able to deliver the current demanded by the Miller effect during turn-on and turn-off.

The following equations may be used to calculate the current flowing in each buffer transistor at turn-on (Q_1) and turn-off (Q_2). The charge current I_{charge} is given by

$$I_{\text{charge}} = \frac{C_{GS}V_{GS}}{t_r} \qquad (4\text{-}30)$$

and

$$C_{GS} = C_{iss} - C_{rss} \qquad (4\text{-}31)$$

where C_{GS} = gate-to-source capacitance, pF
$\quad C_{iss}$ = input capacitance, pF
$\quad C_{rss}$ = reverse transfer capacitance, pF
$\quad V_{GS}$ = gate-to-source voltage, V
$\quad t_r$ = input pulse rise time, ns

If we assume that the gate-to-drain capacitance discharges at the same time, then $t_r = t_f$, and the discharge current is given by the following equation:

$$I_{\text{dis}} = \frac{C_{rss}V_{DS}}{t_r} \qquad (4\text{-}32)$$

where V_{DS} is the drain-to-source voltage (in volts).

In order to calculate the power dissipated in each of the buffer transistors, the following formula is used:

$$P = V_{CE}I_C t_r f \qquad (4\text{-}33)$$

where V_{CE} = buffer transistor saturation voltage, V
$\quad I_C$ = buffer transistor collector current, A
$\quad f$ = transistor switching frequency, kHz

Another way of driving a MOSFET, instead of with discrete transistors, is using special integrated buffers such as the one shown in Fig. 4-26, i.e., the DS0026 high-current driver.

Driving the MOSFET from CMOS   Because of the MOSFET high input impedance, it may be directly driven by a CMOS gate, as shown in Fig. 4-27a. This configuration will produce rise and fall times of about 60 ns. In order to achieve faster switching times, an emitter-follower buffer may be used, as shown in Fig. 4-25, or more than one CMOS gate may be paralleled, as shown in Fig. 4-27b, to increase current availability to the MOSFET input capacitances.

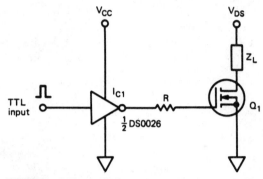

FIGURE 4-26 A high current integrated buffer (i.e., DS0026) may be used to interface TTL levels to a MOSFET, thus considerably improving the switching times.

Driving the MOSFET from Linear Circuits It is conceivable that the MOSFET may be driven directly from the output of an operational amplifier (op-amp), such as a power op-amp capable of delivering high output current. The limiting factor, however, is the slow slew rate of a power op-amp, which limits the operating bandwidth to less than 25 kHz.

In order to improve both bandwidth and slew rate to make the op-amp usable in driving the MOSFET, an emitter-follower buffer may be used. A typical driving circuit using an op-amp is shown in Fig. 4-28.

Some Other Driving Considerations In all the previous driving circuits the MOSFET was used in the common-source configuration. There are times, however, where the common-drain configuration may have to be used, for example, in a bridge circuit. Such an instance requires a totem-pole circuit and thus driving becomes more difficult. The difficulty arises from the fact that as the voltage across the load increases, the enhancement voltage of the common-drain MOSFET decreases.

This can easily be seen from the circuit of Fig. 4-29. In this configuration, when MOSFET Q_3 turns on, the voltage across Z_L rises to the voltage V_2. This means that the enhancement voltage of Q_3 decreases, and unless $V_1 > V_2$, the voltage across Z_L never reaches V_2. It will be necessary, therefore, to produce a voltage at the gate of Q_3 which will be greater than the voltage across the load, and if such a supply is not available, a bootstrap circuit such as the one shown in Fig. 4-30 may be used.

In this circuit, when Q_1 and Q_3 are on, capacitor C is charged to a voltage $(V - V_D)$ through diode D. When Q_1 and Q_3 are turned off, the gate voltage of Q_2 is pulled to the above voltage, and Q_2 turns on, impressing a voltage $(V - V_{GS})$ across Z_L. Of course, since the input impedance of Q_2 is very

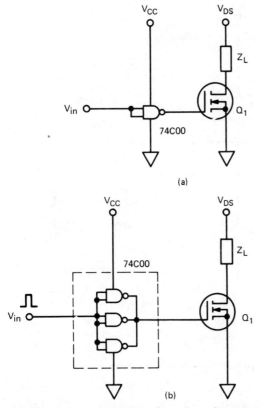

FIGURE 4-27   Circuit (*a*) shows direct drive of a MOS-
FET from a CMOS gate. In order to improve the speed
of the MOSFET, more than one CMOS gate may be
paralleled in order to provide larger gate current (*b*).

high, the charge across C is maintained long enough to turn Q_2 completely
on. Capacitor C must be made large enough to sustain this charge, and a
good first approximation is to choose $C \geq 10C_{iss}$.

Another method used to drive a common-drain MOSFET is the transformer coupled drive. A typical circuit is shown in Fig. 4-31, and it may be
used for bridge circuit designs. The input drive pulses V_{in} are in phase.
Transistor pairs Q_1-Q_2 and Q_5-Q_6 constitute emitter-follower drivers. The
upper MOSFET Q_3 is driven through a transformer, while the lower
MOSFET Q_4 is directly coupled to its drive. Because of the polarity
in the windings of T_1, depicted by the dots, Q_3 is on when Q_4 is off, and vice
versa. Resistors R_1 and R_3 are used to suppress parasitic oscillations, while
resistors R_2 and R_4 are used to assist MOSFET turn-off.

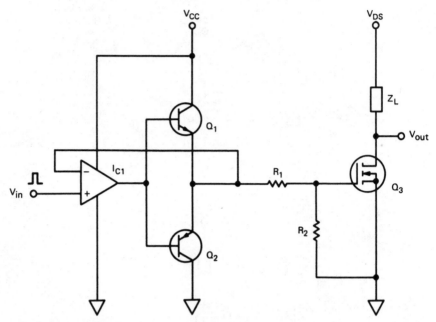

FIGURE 4-28 A typical application of a single-supply op-amp–emitter-follower drive circuit used to drive a MOSFET.

4-9-8 Power MOSFET Switch Protection Circuits

It was mentioned during the presentation of the MOSFET SOA character-istics that it is conceivable to switch maximum power through a power MOSFET without the need of snubbers. Although this statement may be true, it is still a good design consideration to add an RC snubber network across a MOSFET switch. There are basically two main reasons why this is advisable. First, the RC snubber alters the MOSFET load line, increasing its reliability to a maximum, and second the snubber dissipates the excess turn-off energy which otherwise would be dissipated by the transistor switch. Thus transistor stress is minimized without affecting the overall efficiency of the switch.

It is very interesting to note also that by using power MOSFETs, the leakage inductance commutating diode, which is used across the switch to return inductive energy back to the supply bus, is not necessary any more. This is because all MOSFET structures have a body-drain PN junction in shunt with the channel. Figure 4-32 shows a MOSFET used as a switch, depicting the integrated body-drain diode, and also an RC snubber.

Although the transition time for MOSFETs is one order of magnitude less than for bipolars, all the equations presented in Sec. 4-8 for calculating the R and C are valid for the power MOSFET also.

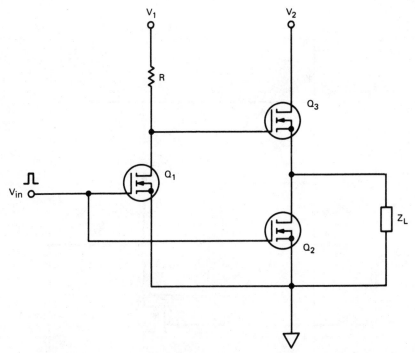

FIGURE 4-29 A totem-pole circuit used to drive a load connected to ground. In this case the MOSFET is operated in the common-drain mode.

4-10 THE GATE TURN-OFF (GTO) SWITCH

The gate turn-off (GTO) semiconductor switch is gaining popularity in switching circuits, especially in equipment which operates directly from European mains. The GTO offers the following advantages over a bipolar transistor: high blocking voltage capabilities, in excess of 1500 V, and also high over-current capabilities. It also exhibits low gate currents, fast and efficient turn-off, as well as outstanding static and dynamic dv/dt capabilities.

Figure 4-33 depicts the symbol of a GTO. The two-transistor circuit is used to explain how the GTO works. When positive drive is applied to the G terminal, transistor Q_2 turns on, driving the base of transistor Q_1 low. Thus, transistor Q_1 turns on allowing collector current to flow into the base of Q_2, setting up regenerative conditions, that is, if the sum of the transistor gains exceeds unity, the device will latch in the on state. Unlike a thyristor though, the GTO will turn off only when a negative gate drive is applied. When the GTO is turned on, its saturation voltage is much lower than bipolar transistors even at high current due to its double-injection manufacturing process.

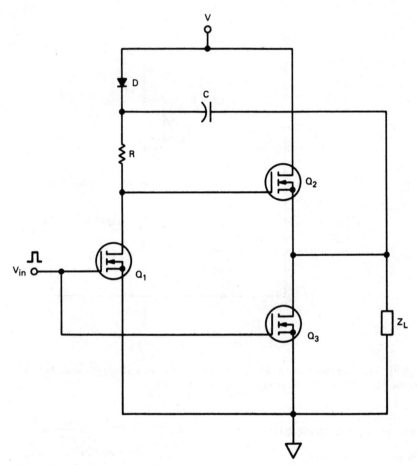

FIGURE 4-30 A bootstrap circuit is used to improve the totem-pole configuration when the MOSFET is operated in the common drain mode.

4-10-1 Gate Drive Requirements of the GTO

Unlike the bipolar transistor, the GTO may be turned on, at comparable applications, with much lower gate drive. Thus, gate drive for a GTO may become very simple, and Fig. 4-34 shows a basic drive circuit.

The GTO is turned on by applying a positive gate current, and it is turned off by applying negative gate-cathode voltage. A practical implementation of a GTO gate drive circuit is shown in Fig. 4-35. In this circuit when transistor Q_2 is off, emitter follower transistor Q_1 acts as a current source pumping current into the gate of the GTO through a 12-V zener Z_1 and polarized

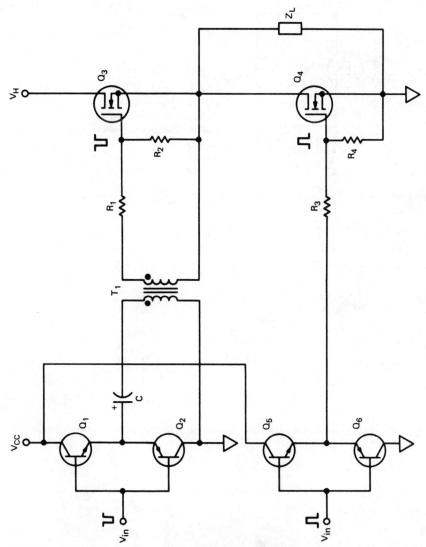

FIGURE 4-31 A transformer coupled common-drain MOSFET configuration. Transformer T_1 is used to enhance the turn-on and turn-off of switching transistor Q_3 without the need of bootstrapping. Gate resistors R_1 and R_3 are parasitic oscillation suppressors, and they must be physically located very close to the MOSFET gates. Resistors R_3 and R_4 assist MOSFET turn-off.

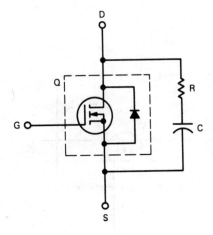

FIGURE 4-32 A power MOS-FET used as a switch, showing the integrated commutating diode. The *RC* snubber network is used to keep the transistor voltage below the breakdown drain-to-source voltage ($V_{B,DS}$).

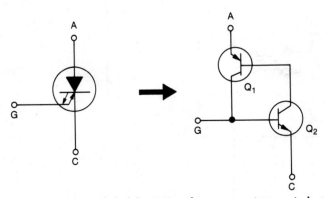

FIGURE 4-33 Symbol of the GTO and its two-transistor equivalent circuit.

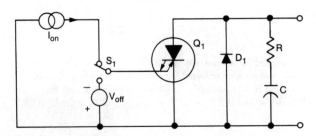

FIGURE 4-34 Basic drive circuit of a GTO.

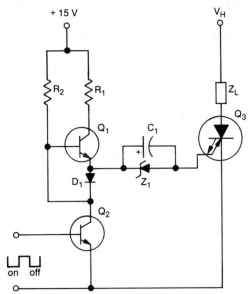

FIGURE 4-35 Practical realization of a GTO gate drive circuit.

capacitor C_1. When the control voltage at the base of Q_2 goes positive, transistor Q_2 turns on, while transistor Q_1 turns off since its base now is one diode drop more negative than its emitter. At this stage the positive side of capacitor C_1 is essentially grounded, and C_1 will act as a voltage source of approximately 10 V, turning the GTO off.

Isolated gate drive circuits may also be easily implemented to drive the GTO.

REFERENCES

1. Bailey, B.: "Reverse Bias Safe Operating Area," Motorola, AN-785, 1978.

2. Clemente, S.: "Gate Drive Characteristics and Requirements for Power HEXFETs," International Rectifier, AN-937, 1980.

3. Hetterscheid, W.: "Base Circuit Design for High-Voltage Switching Transistors in Power Converters," Amperex Electronic Corp., 1974.

4. Hnatek, E. R.: "Design of Solid State Power Supplies," 2d ed., Van Nostrand Reinhold, New York, 1981.

5. Hoffman, D.: "Driving MOSPOWER FETs," Siliconix, AN-79-4, 1979.

6. Oxner, E.: "What are MOSPOWER FETs?" Siliconix Design Catalog, 1982.

7. Pelly, B. R.: "Using High Voltage Power MOSFETs in Off-Line Converter Applications," Powercon 7 Proceedings, 1979.

8. Pressman, A. I.: "Switching and Linear Power Supplies, Power Converter Design," Hayden, Rochelle Park, NJ, 1977.

9. Roark, D.: "Base Drive Considerations in High Power Switching Transistors," TRW, AN-120, 1975.

10. Skanadore, W. R.: "Methods for Utilizing High Speed Switching in High Energy Switching Environments," General Semiconductor Industries, 1978.

11. Walker, R. J.: "Circuit Techniques for Optimizing High Power Transistor Switching Efficiency," Powercon 5, 1978.

12. Wood, P. N.: "Design Considerations for Transistor Converters," TRW, AN-142, 1977.

13. ———: "Switching Power Converters," Van Nostrand Reinhold, New York, 1981.

FIVE

THE HIGH-FREQUENCY POWER TRANSFORMER

5-0 INTRODUCTION

Many engineers consider the design of magnetic components as a sort of "black art." Nothing could be more wrong. The design of magnetic components is an exact science, and it follows precisely all the fundamental electromagnetic laws developed by the pioneer scientists of the field, such as Maxwell, Ampere, Oersted, and Gauss.

The purpose of this chapter is to introduce the fundamental laws of magnetism and to present in a simple, logical, and coherent way the relationships that exist between magnetism and electricity for the design of practical electromagnetic components, such as coils and transformers.

5-1 PRINCIPLES OF ELECTROMAGNETISM

Consider the simple electrical circuit depicted in Fig. 5-1, consisting of a voltage source V, a switch S, and a load L, in the form of an air coil. If at some instance switch S is closed, a current I will flow through the wires to the load. As the current passes through the coil, a magnetic field is established, as shown, which links the turns of the coil. This is called *flux*, and the lines of field are called *flux linkages*.

The flux in this coil, however, is still not very strong. If we were to place a bar of magnetic material (ferromagnetic) inside the coil, as shown in Fig. 5-2, an additional magnetic field would be induced in the bar, producing much more flux. The flux linkages would travel through the bar, and they would find a return path through the surrounding air. If the ferromagnetic core were constructed in such a way to present a continuous path to the flux, then the field would be confined within the core, as shown in Fig. 5-3, inducing a strong magnetic field.

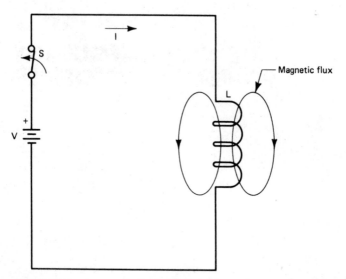

FIGURE 5-1 A magnetic flux is developed in an air-core coil resulting from a current flow *I*.

The degree to which the flux is concentrated is called the *magnetic flux density* or *magnetic induction,* measured at a given point, and designated by the symbol *B*. The units of *B* in the centimeter-gram-second (cgs) system, which is used throughout this book, are given in gauss (G). On the other hand, the magnetizing force that produces the magnetic flux is known as the *magnetic field strength H,* and the units are given in oersteds (Oe).

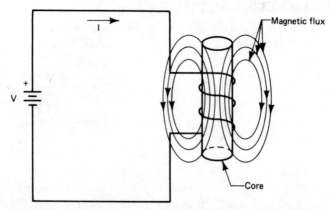

FIGURE 5-2 A bar of ferromagnetic material placed inside the coil will create an additional, stronger flux.

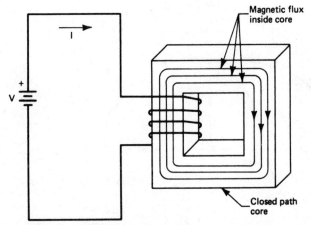

FIGURE 5-3 A continuous ferromagnetic core will confine all the flux within, producing a strong magnetic field.

The magnetic field strength can be written as

$$H = \frac{0.4\pi NI}{l_i} \tag{5-1}$$

where N = coil number of turns
I = magnitude of current flow
l_i = magnetic length of core

Another important relationship between the magnetic flux and the magnetizing force is their ratio, called the *permeability* μ, given as

$$\mu = \frac{B}{H} \tag{5-2}$$

Permeability depicts the ease with which a core material may be magnetized by a certain induction force. The permeability of air is constant and has the value of 1 in the cgs system.

5-2 THE HYSTERESIS LOOP

Every magnetic material is characterized by an S-shaped curve, known as the hysteresis loop. This loop is a curve plotted on *B-H* coordinates by subjecting the magnetic material to a complete magnetizing and demagnetizing cycle. Figure 5-4 shows a typical hysteresis curve of a ferromagnetic core without an air gap in the magnetic flux path. Thus, if we start at point *a* on the curve, which denotes the maximum positive magnetizing force, to point *b*, which denotes zero magnetizing force, then down to point *c*, the

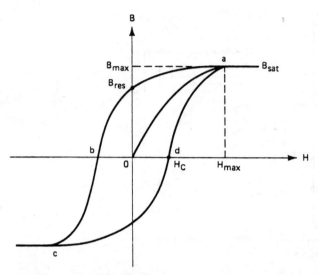

FIGURE 5-4 Hysteresis loop of a typical magnetic core (gapless).

maximum negative magnetizing force, back through zero point d to maximum positive magnetizing force at point a, a complete magnetic cycle is obtained in the form of an S.

Figure 5-4 also depicts certain points on the hysteresis loop that are very important and are defined as follows:

B_{max} is the maximum flux density (induction) point.

H_{max} is the maximum magnetizing force.

B_{res} is the residual magnetic flux (induction), when the magnetizing force is zero.

H_c is the coercive force or the reverse magnetizing force necessary to reduce the residual induction to zero.

It is apparent from the B-H curve of Fig. 5-4 that B_{max} is reached at a certain H_{max} value and that it cannot be exceeded, even if the magnetizing force is increased. The value of the magnetic induction at this point is said to be saturated and is written as B_{sat}.

If we were to introduce an air gap into the core, a hybrid flux path would be created, which would alter the effective length of the magnetic path. Since the air gap permeability is unity, the effective magnetic path length is

$$l_e = l_i + \mu_i l_g \tag{5-3}$$

where l_i = length of magnetic path of material
l_g = length of magnetic path of air gap
μ_i = permeability of magnetic material

Applying Ampere's circuital law to the gapped core, one can show that the core flux density may be written as

$$B_i = \frac{0.4\pi NI\mu_i}{l_i + \mu_i l_g} \tag{5-4}$$

Equation 5-4 is a very important relationship, since it states that for a given ampere-turn product (NI), the flux density of a core with an air gap is smaller than that of a gapless core. In other words, B_{sat} with an air gap is smaller than B_{sat} without an air gap. Thus the introduction of an air gap in a magnetic circuit gives a "tilt" to the hysteresis loop, as shown in Fig. 5-5, reducing the possibility of core saturation at high magnetizing force.

The majority of magnetic core manufacturers describe the B-H property of their material by the normal magnetizing curve, as depicted in Fig. 5-6. This curve shows that the slope of B vs. H in the region below the "knee" may be considered essentially constant. Thus a linear relationship exists at this region between the excitation current and the resultant flux, which makes the permeability of the core also constant.

In the low-level region of the curve, core losses become negligible; thus the core temperature stays low. Above the knee the core goes into saturation, and operation at this region must be avoided for linear applications.

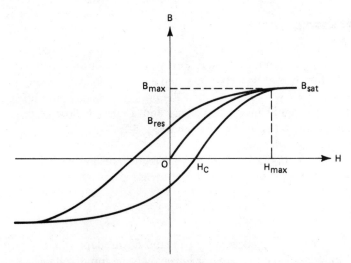

FIGURE 5-5 A hysteresis loop of a core with an air gap. Notice the shearing of the loop due to the addition of the air gap flux and the reduction of the B_{sat} value compared to the gapless case.

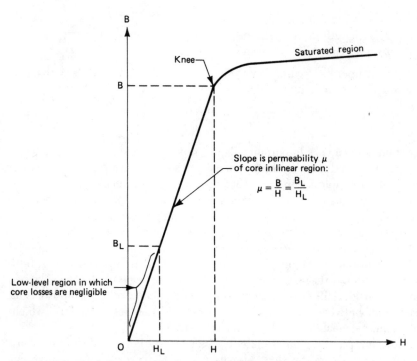

FIGURE 5-6 Typical magnetization curve showing linear and saturation regions.

5-3 BASIC TRANSFORMER THEORY

We mentioned in previous discussions that a current flowing through the windings of a coil wound around a closed core will induce a flux confined within the core. If this current were periodic and a second coil were wound around the same core, we would expect that the opposite effect would take place; that is, the flux would induce a voltage and a current flow in this secondary winding. Indeed this is the case, and Fig. 5-7 shows the simplest form of a two-winding transformer.

Normally a transformer operates with high efficiency in raising or lowering the output voltage in proportion to the ratio of turns, given by

$$\frac{N_P}{N_S} = \frac{V_P}{V_S} \tag{5-5}$$

Thus, transformers are categorized as step-up or step-down, depending upon whether the secondary voltage is higher or lower than the input voltage. Of course more than one secondary winding may be incorporated, which could produce both higher and lower voltages. One of the most important and

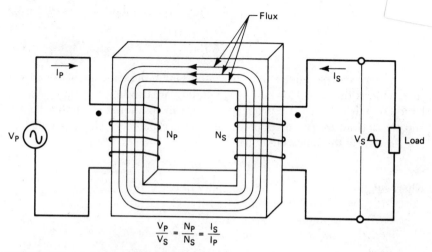

$$\frac{V_P}{V_S} = \frac{N_P}{N_S} = \frac{I_S}{I_P}$$

FIGURE 5-7   Typical two-winding transformer.

useful characteristics of a transformer is the electrical isolation which it offers between primary and secondary.

From the fundamental transformer magnetic relationship

$$e = NA_e \left(\frac{dB}{dt}\right) 10^{-8} \tag{5-6}$$

we can derive an expression to calculate the flux density B in order to make sure that the transformer operates in the linear portion of the magnetization curve. This expression is given by

$$B_{max} = \frac{(V_P)10^8}{KfN_PA_e} \tag{5-7}$$

where V_P = impressed primary voltage, V
$\quad f$ = frequency, Hz
$\quad N_P$ = primary number of turns
$\quad A_e$ = core effective area, cm^2
$\quad K$ = 4.44 for sine waves

Since this book is devoted to switching power supplies, $K = 4$ will be used for push-pull and bridge converters, and $K = 2$ will be used for forward converters.

Normally, the designer of the power transformer chooses B_{max} arbitrarily, so that it lies within the linear region of the $B\text{-}H$ curve. A good starting point is to choose $B_{max} = B_{sat}/2.$

A simple rearrangement of Eq. 5-7 yields the number of primary turns

$$N_P = \frac{(V_P)10^8}{KfB_{max}A_e} \qquad (5\text{-}8)$$

Two more design parameters are very important in the selection of the proper core. One is the core's (or bobbin's) winding area, which must be selected large enough to carry the proper wire size in minimizing winding losses, and the second is the core's power handling capability. These parameters are related by the following equation:

$$P_{out} = (1.16B_{max}fdA_eA_c)10^{-9} \qquad (5\text{-}9)$$

where P_{out} = power handling capability of core, W
B_{max} = peak operating flux density, G
f = frequency, Hz
d = current density of wires, A/m^2
A_e = core effective area, cm^2
A_c = bobbin winding area, cm^2

Some manufacturers use the symbol W_a for window area instead of A_c. Normally the current density is expressed in circular mils per ampere, symbolized as D, and related to d by

$$d = \frac{1.27 \times 10^6}{D} \qquad (5\text{-}10)$$

Substituting Eq. 5-10 into Eq. 5-9 we get

$$P_{out} = \frac{(1.47fB_{max}A_eA_c)10^{-3}}{D} \qquad (5\text{-}11)$$

Rearranging Eq. 5-11 we derive a very useful equation for calculating and selecting the core size of a transformer:

$$A_eA_c = \frac{(0.68P_{out}D)10^3}{fB_{max}} \text{ cm}^4 \qquad (5.12)$$

Operating current density D is given by the wire manufacturer based on 1000 circular mils per ampere (c.m./A). Practical designs use a current density below that number, and depending upon the application and the number of the winding turns, current densities as low as 200 c.m./A may be used safely.

5-4 CORE MATERIAL AND GEOMETRY SELECTION

Although almost any magnetic material may be used in designing high-frequency power transformers, ferrites have been almost exclusively used in modern converter designs. Ferrites may not have very high operating flux densities—most ferrites have a B_{sat} from 3000 to 5000 G—but they offer low core losses at high frequencies, good winding coupling, and ease of assembly.

Cores made from ferrites come in many shapes and sizes, and various power ferrite materials specifically aimed at high-frequency transformer design have been developed by manufacturers. Table 5-1 names some of the most popular ferrite trade materials and their manufacturer.

The geometry of the core used for a specific application depends on the power requirements. E-E, E-I, E-C, and pot cores are some of the most popular shapes. Because of their construction, pot cores are very well suited for low- to medium-power applications, anywhere from 20 to 200 W. They are particularly attractive for designs requiring low flux leakage, and their inherent self-shielding design minimizes EMI.

For higher power levels E-E, E-I, and E-C cores may be used. The E-C core is a compromise between an E-E and a pot core, combining the advantages of each shape.

Manufacturers of cores list all the important parameters required to design a power transformer, and if a particular parameter is not listed, it can easily be calculated using the equations in Sec. 5-3. Table 5-2 relates magnet wire size, AWG (American wire gauge), to current densities.

5-5 DESIGN OF A POWER TRANSFORMER FOR A PULSE-WIDTH-MODULATED HALF-BRIDGE CONVERTER

The step-by-step design example which follows demonstrates the development of a typical high-frequency power transformer. The example is general, and judicial changes will make it useful for full-bridge or push-pull pulse-width-modulated (PWM) power converter designs. The materials chosen for the design were selected as a representative sample, but any material could

TABLE 5-1 FERRITE CORE MATERIAL FOR HIGH-FREQUENCY POWER TRANSFORMERS

Material	Manufacturer
3C8	Ferroxcube
24B	Stackpole
77	Fair-Rite Products
F, T	Magnetics, Inc.
H7C1	TDK
N27	Siemens

TABLE 5-2   HEAVY FILM-INSULATED MAGNET WIRE SPECIFICATIONS

| AWG | Diameter over insulation (inches) | | Nominal circular mil area | Resistance per 1000 ft | Current capacity in milliamperes based on 1000 c.m./A | AWG |
	Min.	Max.				
8	0.130	0.133	16510	0.6281	16510	8
9	0.116	0.119	13090	0.7925	13090	9
10	0.104	0.106	10380	0.9985	10380	10
11	0.0928	0.0948	8230	1.261	8226	11
12	0.0829	0.0847	6530	1.588	6529	12
13	0.0741	0.0757	5180	2.001	5184	13
14	0.0667	0.0682	4110	2.524	4109	14
15	0.0595	0.0609	3260	3.181	3260	15
16	0.0532	0.0545	2580	4.020	2581	16
17	0.0476	0.0488	2050	5.054	2052	17
18	0.0425	0.0437	1620	6.386	1624	18
19	0.0380	0.0391	1290	8.046	1289	19
20	0.0340	0.0351	1020	10.13	1024	20
21	0.0302	0.0314	812	12.77	812.3	21
22	0.0271	0.0281	640	16.20	640.1	22
23	0.0244	0.0253	511	20.30	510.8	23
24	0.0218	0.0227	404	25.67	404	24
25	0.0195	0.0203	320	32.37	320.4	25
26	0.0174	0.0182	253	41.02	252.8	26
27	0.0157	0.0164	202	51.44	201.6	27
28	0.0141	0.0147	159	65.31	158.8	28
29	0.0127	0.0133	128	81.21	127.7	29
30	0.0113	0.0119	100	103.7	100	30
31	0.0101	0.0108	79.2	130.9	79.21	31
32	0.0091	0.0098	64	162	64	32
33	0.0081	0.0088	50.4	205.7	50.41	33
34	0.0072	0.0078	39.7	261.3	39.69	34
35	0.0064	0.0070	31.4	330.7	31.36	35

be used as long as the manufacturer's data sheets are consulted for use of the proper specifications.

DESIGN EXAMPLE 5-1

Design a 100-W power transformer to be used in a 20-kHz half-bridge, PWM converter which operates from 90- to 130- and 180- to 260-V ac mains and produces 5 V at 20 A output.

DESIGN PROCEDURE

Step 1: Choose the core geometry and ferrite material. For this design choose a Ferroxcube pot-core ferrite of 3C8 material. Use the circuit of Fig. 3-12 as a guide for the design.

Step 2: Choose a working B_{max}. From the Ferroxcube catalog specifications for the 3C8 material we find that the saturation flux density at 100°C is B_{sat} = 3300 G. Since the converter has to work from 90- to 130- and 180- to 260-V ac, we take B_{max} at 90 V ac to be 1600 G. This choice guarantees that B_{max} will stay below 3300 G at 130 V ac, thus the transformer will not saturate.

Step 3: Find the maximum working primary current. The transformer primary has to conduct the maximum possible current at the low input voltage of 90 V ac. The dc voltage after the rectifiers is V_{in} = 2(90 × 1.4) = 252 V. Using Eq. 3-28, the primary current is

$$I_P = \frac{3P_{out}}{V_{in}} = \frac{3 \times 100}{252} = 1.19 \text{ A}$$

Step 4: Determine core and bobbin size. We choose to work with a current density of 400 c.m./A. Use Eq. 5-12 to calculate the $A_e A_c$ product

$$A_e A_c = \frac{0.68 \times 100 \times 400 \times 10^3}{20 \times 10^3 \times 1600} = 0.850 \text{ cm}^4$$

Choose a core size close to the calculated $A_e A_c$ product of 0.850 cm^4. The 2616 core has A_e = 0.948, but from the same catalog the single-section bobbin has a winding area A_c = 0.406, giving an $A_e A_c$ product value of 0.384 cm^4, which is too low.

Try the 3019-PL00-3C8 core and bobbin. From the manufacturer's data sheets we get A_e = 1.38 cm^2 and A_c = 0.587 cm^2. Then $A_e A_c$ = 0.810 cm^4, which is very close to the calculated value of 0.850 cm^4.

Although this core is able to handle the required power, it is a good practice to choose an $A_e A_c$ value which is at least 50 percent higher than the calculated one in order to account for insulation thickness and air space around round wires. Thus for this design we choose the 3622-PL00-3C8 Ferroxcube pot core and the 3622F1D bobbin. From the manufacturer's data we get A_e = 2.02 cm^2 and A_c = 0.748 cm^2, yielding $A_e A_c$ = 1.5 cm^4. This value is more than adequate for the transformer design.

Step 5: Calculate wire size and primary number of turns. Since we chose 400 c.m./A to be the wire current density for this design, the primary winding requires a wire size of $1.19 \times 400 = 476$ c.m., which from the magnet wire specifications of Table 5-2 corresponds to no. 23 AWG.

From the Ferroxcube catalog, the turns per bobbin graph of the 3622F1D single-section bobbin shows that approximately 180 turns of no. 23 wire are required to fill the bobbin. Assuming that the primary winding fills 30 percent of the bobbin winding area, if the primary turns are calculated to be 60 turns or less, then the core and bobbin choice is correct.

Again taking the worst-case operating condition of 90 V ac, $V_{in,min} = 90 \times 1.4 - 20$ V dc ripple and rectifier drop $= 107$ V dc. Using Eq. 5-8 the primary number of turns is calculated.

$$N_P = \frac{107 \times 10^8}{4 \times 1600 \times 20 \times 10^3 \times 2.02} = 41.3 \text{ turns}$$

We round off the number of turns to 40, which is below the theoretical value of 60; therefore the core and bobbin choice is a good one.

Step 6: Check B_{max} at $V_{in,max}$. Using the calculated number of turns, we are now able to calculate the maximum working flux density of the transformer at $V_{in,max} = 130 \times 1.4 + 20$ V dc for ripple voltage $= 202$ V dc. Using Eq. 5-8, solving for B_{max},

$$B_{max} = \frac{202 \times 10^8}{4 \times 40 \times 20 \times 10^3 \times 2.02} = 3125 \text{ G}$$

The value of 3125 G is below the specified saturation flux density of the Ferroxcube 3C8 material, which is specified as $B_{sat} \geq 4400$ G at 25°C and $B_{sat} \geq 3300$ G at 100°C. If a greater B_{sat} margin is required, then the value of B_{max} used in step 5 must be chosen below 1600 G.

Step 7: Calculate the number of layers used by the primary winding. From Table 5-2, we find that no. 23 AWG has a maximum diameter of 0.025 in. for double-insulation wire. The Ferroxcube catalog gives the bobbin window width as 0.509 in. Therefore the maximum number of turns per layer, using no. 23 AWG, is $0.509/0.025 = 20.4$ turns. Thus the primary winding occupies two layers, 20 turns per layer.

Step 8: Calculate the transformer secondary turns. Because the output voltage is derived from a full-wave center-tap rectifier scheme using PWM technique, then $V_S = 2V_{out}$ at minimum V_{in}, where V_{out} is the nominal output voltage and 2 is the factor that averages out the approximate 50 percent duty cycle. Since we want to maintain output voltage regulation at $V_{in,min}$, the number of secondary turns required is

$$N_S = N_P \frac{V_S}{V_P} = 40 \frac{10}{107} = \frac{400}{107} = 3.74 \text{ turns}$$

We will use 4 turns in the secondary.

Step 9: Calculate the secondary winding wire size and number of layers. We have already mentioned the fact that the secondary is using a full-wave center-tap rectification scheme, therefore each secondary half conducts approximately 50 percent of the load current, or 10 A. Taking the current density at 400 c.m./A, then for each half of the secondary winding we need 400 c.m./A × 10 A = 4000 c.m., which corresponds to no. 14 AWG. In order to minimize copper losses due to skin effects, it is advisable to use lower gauge paired conductors for each winding half, or four wires of 200 c.m. each, for the entire secondary.

At 2000 c.m., we can use no. 17 AWG wire, which has a maximum diameter of 0.049 in. Then, for the entire secondary, the number of turns per layer is 0.509/4(0.049) = 2.69 turns. Therefore the entire secondary of 4 turns occupies two layers.

Step 10: Check for fit. From the Ferroxcube catalog, the height of the 3019F1D bobbin window is calculated to be approximately 0.260 in. From steps 7 and 9, the two windings stack up to a height of 2(0.025) + 2(0.049) = 0.148 in. Assuming that an additional 0.010 in. of tape is used for insulation and finish, the total height is approximately 0.160 in., which is below the available 0.260 in.; therefore the bobbin will comfortably accept all the transformer windings.

5-6 PRACTICAL CONSIDERATIONS

When testing the transformer in the actual application, some fine tuning may be required to better its overall performance. Although most transformers are manufactured stacking one winding on top of the other, as shown in Fig. 5-8a, interleaving the windings may be necessary to reduce the effects of leakage inductance. Interleaving is done by winding half the secondary, followed by the primary winding, followed by the other secondary half, as shown in Fig. 5-8b.

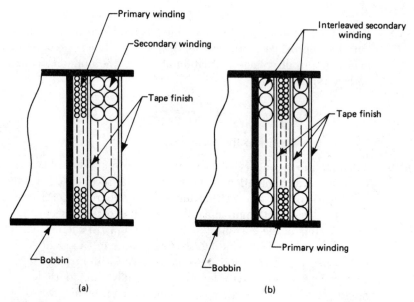

FIGURE 5-8 (*a*) A standard transformer construction with stacked windings; (*b*) the same transformer in an interleaved configuration, where the primary is sandwiched between a split secondary winding.

Some designs may also require a Faraday shield between the primary and secondary to reduce radio frequency interference (RFI) radiation, although pot cores exhibit excellent shelf-shielding properties, since all the windings are covered by the core material. Whatever the needs are, a good transformer design enhances the operation of the power supply in more than one way, and care must be taken during its design.

5-7 THE FLYBACK CONVERTER'S TRANSFORMER-CHOKE DESIGN

In Chap. 3 the basic operation of the flyback converter was described, and Fig. 3-4 depicted some of the fundamental waveforms associated with this converter. Since the isolation element of this topology has the dual function of both transformer and choke, the term transformer-choke has been used to indicate its usage.

In the flyback converter two modes of operation are possible for the transformer-choke: (1) complete energy transfer, where all the energy stored in the inductor-transformer is transferred to the secondary before the transistor switch is turned on and (2) incomplete energy transfer, where not all the energy stored in the transformer-inductor is transferred to the secondary before the transistor switch is turned on. Figure 5-9 shows the waveforms of the two modes of operation.

The complete energy transfer waveforms show a high peak collector current during the turn-on period of the switching transistor. This means that a relatively low primary inductance value is needed to achieve this current rise, at the expense of increased winding losses and input capacitor ripple current. Also, the switching transistor must have a high current carrying capacity to sustain the peak current.

The incomplete energy transfer mode, on the other hand, exhibits a relatively lower peak switching transistor collector current at the expense of a higher collector current flow as the transistor switches on, a fact which can lead to high transistor dissipation. However, since a relatively high transformer-choke primary inductance is needed to achieve this mode of operation, the residual stored energy in the transformer core assumes that the volume of an incomplete energy transfer transformer-choke will have a larger volume than the complete energy transfer one, all other factors being equal.

5-7-1 Design Procedure

In the following steps the necessary equations are given to design a flyback converter transformer-choke for a complete energy transfer mode. An incomplete energy transfer transformer-choke design follows the basic steps

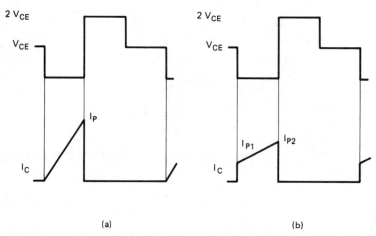

(a)   (b)

FIGURE 5-9 (a) The waveforms depict the voltage and current relationship of a complete energy transfer flyback transformer-choke; (b) the waveforms depict those of an incomplete energy transfer mode.

given below, with some minor changes in the definition of peak collector current (Fig. 5-9b), which may be written as $(I_{P1} - I_{P2})$.

Step 1: Transformer peak primary current. It is necessary to calculate the transformer peak primary current first, which is also equal to transistor peak collector current. From the fundamental inductor voltage relationship, the rate of rise is determined by

$$V = L\frac{di}{dt} \tag{5-13}$$

Since in the complete energy transfer mode, the current ramps up from zero to peak collector current during the time t_C when the switch is closed, the input voltage may be written as

$$V_{in} = L_P \frac{I_{PP}}{t_C} \tag{5-14}$$

Taking $1/t_C = f/\delta_{max}$, then Eq. 5-14 becomes

$$V_{in,min} = \frac{L_p I_{PP} f}{s} \tag{5-15}$$

where V_{in} = dc input voltage, V
L_P = transformer primary inductance, mH
I_{PP} = peak transformer currents, A
δ_{max} = maximum duty cycle, μs
f = switching frequency, kHz

The output power in the complete energy transfer mode is equal to the energy stored per cycle times the operation frequency.

$$P_{out} = \tfrac{1}{2} L_P I_{PP}^2 f \tag{5-16}$$

Dividing Eq. 5-16 into Eq. 5-15 we get

$$\frac{P_{out}}{V_{in,min}} = \frac{L_P I_{PP}^2 f \delta_{max}}{2 L_P I_{PP} f}$$

which yields, by rearranging terms, the transformer peak primary current

$$I_{PP} = I_C = \frac{2P_{out}}{V_{in,min}\delta_{max}} \tag{5-17}$$

Step 2: Relate minimum and maximum duty cycles. In the flyback converter, regulation is accomplished by varying the duty cycle of the switch through predetermined limits, which are designated as δ_{min}

and δ_{max}. If the converter input voltage varies from $V_{in,min}$ to $V_{in,max}$, then

$$\delta_{min} = \frac{\delta_{max}}{(1 - \delta_{max})K + \delta_{max}} \tag{5-18}$$

where

$$K = \frac{V_{in,max}}{V_{in,min}} \tag{5-19}$$

Step 3: Calculate transformer primary inductance. Since peak primary current is now known, the transformer-choke primary inductance may be calculated by

$$L_P = \frac{V_{in,min}\delta_{max}}{I_{PP}f} \tag{5-20}$$

Step 4: Select the minimum size core. From a magnetic core catalog pick the core material and geometry which best suit your application. If we were to wind only the primary winding into a bobbin, the winding area A_c and the core effective area A_e would be related by

$$A_cA_e = \frac{(6.33L_PI_{PP}D^2)10^8}{B_{max}} \tag{5-21}$$

where D is the diameter of the insulated wire (heavy polynylon wire is recommended), and $B_{max} = B_{sat}/2$.

Since we are to design a transformer-choke, a secondary winding(s) will also be associated with the design. Assuming that the primary winding occupies 30 percent of a bobbin available winding area, 70 percent will be reserved for the secondary, for air space around round conductors, and for insulating tape. Therefore the right-hand side of Eq. 5-21 must be multiplied by 3 to take the transformer-choke secondary into account. Experience dictates that an extra safety factor must be added; therefore the multiplication factor will be 4, and Eq. 5-21 becomes

$$A_cA_e = \frac{(25.32L_PI_{PP}D^2)10^8}{B_{max}} \tag{5-22}$$

Of course Eq. 5-22 gives a first-order approximation and the final core and bobbin choice may vary.

Step 5: Calculate the core air gap length. The flyback converter is operating single-endedly; that is, the transformer-choke uses just half the flux capacity, since the current and flux never go negative. This fact may present a potential problem, driving the core into saturation.

To handle the problem two solutions are possible. First, use a core with a very large volume, or second, introduce an air gap in the flux path to flatten the hysteresis loop, thus lowering the working flux density for the same dc bias. Normally, designers perfer the second solution, which offers more compact transformers to be manufactured.

The air gap presents the largest magnetic reluctance in the flux path, and most of the stored energy in the transformer-choke is in the air gap volume v_g, whose length is l_g. Then

$$\tfrac{1}{2} L_P I_{PP}^2 = (\tfrac{1}{2} B_{max} H v_g)10^8$$

where

$$v_g = A_e l_g$$

and

$$\mu_0 H = \frac{B_{max}}{0.4\pi}$$

μ_0 = air permeability = 1. Therefore, the air gap length is given by

$$l_g = \frac{(0.4\pi L_P I_{PP}^2)10^8}{A_e B_{max}^2} \text{ cm} \tag{5-23}$$

If an E-E type core, or similar type, is used to construct the transformer-choke, the center leg may be gapped to the air gap length l_g, or l_g may be equally divided between the outside legs of the core if a spacer is used.

Step 6: Calculate transformer number of primary turns. Knowing the air gap length, the primary number of turns of the transformer-choke may be calculated as follows:

$$N_P = \frac{B_{max} l_g}{0.4\pi I_{PP}} \tag{5-24}$$

The same result may also be derived by using the following equivalent equation:

$$N_P = \frac{(L_P I_{PP})10^8}{A_e B_{max}} \tag{5-25}$$

Either Eq. 5-24 or 5-25 will give the same results.

Step 7: Calculate number of secondary turns. The secondary voltage V_S must be calculated when the input voltage (i.e., primary voltage) is at its minimum with the duty factor at its maximum. It should also be noted that $V_{in,min} = 1.4V_{in,ac} - 20$ V dc ripple voltage and diode drop.

Taking the output rectifier diode drop into account, the output voltage of a specific secondary winding is written

$$V_{out} + V_D = V_{in,min} \frac{\delta_{max}}{1 - \delta_{max}} \frac{N_S}{N_P}$$

Accordingly,

$$N_S = \frac{N_P(V_P + V_D)(1 - \delta_{max})}{V_{in,min}\delta_{max}} \tag{5-26}$$

DESIGN EXAMPLE 5-2

Design the power transformer to be used in a 100-W, complete energy transfer flyback converter (see Fig. 3-4), delivering 5 V dc at 20 A at the output, and operating over an input voltage range of 90 to 130 V ac.

DESIGN PROCEDURE

Step 1: Calculate the peak primary current. Assume that the converter has a maximum duty factor $\delta_{max} = 0.45$. Since the minimum ac voltage input is 90 V, then $V_{in,min} = 90 \times 1.4 - 20$ V dc for ripple and diode drop $= 107$ V dc. Using Eq. 5-17, the peak primary current is

$$I_{PP} = \frac{2P_{out}}{V_{in,min}\delta_{max}} = \frac{2 \times 100}{107 \times 0.45} = 4.15 \text{ A}$$

A switching transistor which can handle this peak collector current at turn-on must also be used in the design.

Step 2: Find the minimum duty factor δ_{min}. The maximum dc input voltage, after rectification, is

$$V_{in,max} = 130 \text{ V ac} \times 1.4 - 0 \text{ V dc ripple} = 182 \text{ V dc}$$

Allow a 10 percent margin, then $V_{in,max} = 200$ V dc. Also allow a 7 percent margin on the $V_{in,min}$ voltage, then $V_{in,min} = 100$ V dc. Consequently the input voltage ratio K is

$$K = \frac{V_{in,max}}{V_{in,min}} = \frac{200}{100} = 2$$

Using Eq. 5-18

$$\delta_{min} = \frac{\delta_{max}}{(1 - \delta_{max})K + \delta_{max}} = \frac{0.45}{(1 - 0.45)2 + 0.45} = 0.29$$

Therefore the converter will operate over the duty ratio range of $0.29 < \delta < 0.45$ for the input voltage range of 200 V dc $> V_{in} >$ 100 V dc.

Step 3: Calculate transformer primary inductance. Using Eq. 5-20 yields

$$L_P = \frac{V_{in,min}\delta_{max}}{I_{PP}f} = \frac{100 \times 0.45}{4.15 \times 20 \times 10^3} = 0.54 \times 10^{-3} \text{ H}$$

Therefore

$$L_P = 540 \ \mu\text{H}$$

Step 4: Select core and bobbin size. Choose a design current density value of 400 c.m./A for the winding magnetic wires. Then

$$400 \text{ c.m./A} \times 4.15 \text{ A} = 1550 \text{ c.m.}$$

From Table 5-2, the value of 1660 c.m. corresponds approximately to AWG no. 18, which has a diameter of 0.044 in.

We will also choose a Ferroxcube 3C8 material, E-C type core. The 3C8 ferrite material has a $B_{sat} = 3300$ G at 100°C, and for this design $B_{max} = B_{sat}/2 = 3300/2 = 1650$ G. Accordingly,

$$A_c A_e = \frac{(25.32 L_P I_{PP} D^2)10^8}{B_{max}}$$

$$= \frac{25.32 \times 540 \times 10^6 \times 4.15 \times 0.044^2 \times 10^8}{1650} = 6.7 \text{ cm}^4$$

From the Ferroxcube catalog the EC70-3C8 core and 70PTB bobbin yield

$$A_e A_c = 2.79 \times 4.77 = 13.3 \text{ cm}^4$$

This $A_e A_c$ value is of course much higher than the required one, but it is the only E-C core available in the catalog which satisfies the requirement $A_c A_e \geq 6.7 \text{ cm}^4$. We will use this core-bobbin combination for the design.

Step 5: Calculate the air gap length l_g. In order to be able to use an ungapped core the effective core volume v_e listed in the catalog must be equal to or greater than the theoretical v_e value given by

$$v_e = \frac{(0.4\pi)10^8(L_P I_P^2)}{B_{max}H}$$

Since we choose $B_{max} = 1650$ G at 100°C, from the 3C8 material magnetization curves in the Ferroxcube catalog we find $H = 0.4$ Oe. Therefore

$$v_{e,min} = \frac{0.4 \times 3.14 \times 10^8 \times 0.54 \times 10^{-3} \times 4.15^2}{1650 \times 0.5}$$

$$= 1415 \text{ cm}^3$$

This effective ungapped core volume requires a very large core size, since the EC70-3C8 core lists an effective core volume of 18.8 cm³ only. In order to be able to use the EC70-3C8, it must be gapped to a length l_g given by Eq. 5-23:

$$l_g = \frac{(0.4\pi L_P I_{PP}^2)10^8}{A_e B_{max}^2} = \frac{11.66 \times 10^5}{75.96 \times 10^5} = 0.15 \text{ cm}$$

The center leg of the E-C core may be gapped to 0.15 cm, or a spacer 0.075 cm thick may be placed between the outer legs of the core in order to achieve the same gap effect.

Step 6: Transformer primary number of turns. Now that all parameters are known, the primary number of turns to achieve the desired inductance is calculated using Eq. 5-24

$$N_P = \frac{B_{max} l_g}{0.4\pi I_{PP}} = \frac{1650 \times 0.15}{0.4 \times 3.14 \times 4.15} = 47.48 \text{ turns}$$

We will use 48 turns in the primary. Using Eq. 5-28 to calculate the primary number of turns

$$N_P = \frac{(L_P I_{PP})10^8}{A_e B_{max}} = \frac{0.54 \times 10^{-3} \times 4.15 \times 10^8}{2.79 \times 1650} = 47.68 \text{ turns}$$

Thus, both equations give equivalent results.

Step 7: Transformer secondary number of turns. Using Eq. 5-26, we get

$$N_S = \frac{N_P(V_{out} + V_D)(1 - \delta_{max})}{V_{in,min}\delta_{max}} = \frac{48(5 + 1)(1 - 0.45)}{100 \times 0.45}$$

$$= 3.52 \text{ turns}$$

Since there will also be some voltage drops in the printed circuit conductors and the output winding copper conductors which were not taken into account in the above equation, the number of secondary turns may be taken to be $N_S = 4$ turns.

The output voltage on a flyback converter requires a single winding, one diode, and a capacitor, as shown in Fig. 3-4. In order to deliver 20 A of output current at 400 c.m./A, a wire of 20 × 400 = 8000 c.m. is required. To minimize losses due to skin effect, four wires in parallel of 2000 c.m. per wire will be used. This corresponds to using four AWG no. 17 wires in parallel.

There should be no problem in fitting all the windings, plus insulation, in the chosen bobbin because the $A_e A_c$ product was much larger than the calculated one, by a factor of 2.

5-8 SOME GENERAL HIGH-FREQUENCY TRANSFORMER CONSIDERATIONS

In the previous paragraphs, some practical and fundamental design equations and procedures were given for the design of the isolation transformer of certain types of converters. The majority of the design equations are fundamentally valid for any type of magnetic circuit, and they may be adapted to solve a variety of magnetic applications, whether transformers, chokes, or a combination of both.

In general magnetic components used in the construction of a switching power supply have to comply with certain national or international safety standards. Thus, for the North American countries, standards set by the Underwriters Laboratories (UL) for the United States and Canadian Standards Association (CSA) for Canada are valid. For European use, the West German Verband Deutscher Elektronotechniker (VDE) safety standards have become the most popular design guidelines, since they are considered to be the most stringent.

There are some fundamental differences between the UL and VDE safety standards, the UL concentrating more on preventing fire hazards, while the VDE is more concerned with the safety of the operator. In constructing an isolation transformer, UL and CSA limit the winding temperature rise to 65°C above ambient for class 105 insulation and to 85°C above ambient for class 130 insulation.

In any case, it is a good design practice to keep the temperature rise of a switching power supply transformer to a low value, since most of these transformers are constructed using ferrites, which have thermal limitations. Ferrites have a Curie temperature of about 200°C, which limits their operating core temperature to about 100°C. Curie temperature is the temperature at which a material changes its ferromagnetic properties and becomes paramagnetic.

The VDE safety standards, on the other hand, have strict requirements for specific winding techniques and input-to-output isolation requirements, which may require a 3750-V ac hi-pot test potential. An extensive discussion of these safety requirements is given in Chap. 11.

Varnish impregnation may not be necessary with a ferrite transformer, since oxidization of the core due to moisture, as was the case with iron laminations, is not a factor. Also the acoustical noise which was associated with low-frequency transformers is not present in the ferrite high-frequency transformers, which generally operate above the human acoustical range. That is not to say that ferrite transformer assemblies may not generate mechanical or acoustical noise. They may, because whatever they are mounted on acts as a sounding board. A phenomenon associated with ferrites, called magnetostriction, shortens or lengthens the part due to the applied magnetic field, which in turn causes mechanical resonance of the core assembly. In fact magnetostriction changes polarity, from negative to positive, as the temperature of the core rises. Therefore, care must be taken to use proper methods when mounting the transformer to the board, in order to reduce or eliminate any acoustical mechanical noise.

REFERENCES

1. Ferroxcube (a) "Linear Ferrite Materials and Components," Ferroxcube Corp., Magnetic Catalog; (b) "Linear Ferrite Magnetic Design Manual," Ferroxcube Corp., Bulletin 550, 1971.

2. Manka, W. V.: Design Power Inductors Step By Step, *Electronic Design*, vol. 25, no. 26, 1980.

3. Middlebrook, R. D., and S. Ćuk: "Advances in Switched Mode Power Conversion," vols. 1 and 2, Teslaco, Pasadena, California, 1981.

4. Pressman, A. I.: "Switching and Linear Power Supply, Power Converter Design," Hayden, Rochelle Park, NJ, 1977.

5. Thibodeau, P. E.: The Switching Transformer: Designing It in One Try for Switching Power Supplies, *Electronic Design*, vol. 28, no. 18, 1980.

6. Watson, J. K.: "Applications of Magnetism," Wiley, New York, 1980.

SIX

THE OUTPUT SECTION: RECTIFIERS, INDUCTORS, AND CAPACITORS

6-0 INTRODUCTION

In general the output section of any switching power supply is comprised of single or multiple dc voltages, which are derived by direct rectification and filtering of the transformer secondary voltages and in some cases further filtering by series-pass regulators. These outputs are normally low-voltage, direct current, and capable of delivering a certain power level to drive electronic components and circuits. Most common output voltages are 5 V dc, 12 V dc, 15 V dc, 24 V dc, or 28 V dc, and their power capability may vary from a few watts to thousands of watts.

The most common type of secondary voltages that have to be rectified in a switching power supply are high-frequency square waves, which in turn require special components, such as Schottky or fast recovery rectifiers, low ESR capacitors, and energy storage inductors, in order to produce low noise outputs useful to the majority of electronic components.

This chapter describes the characteristics, merits, and limitations of the components used in the output section of the switching power supply. Design equations and procedures are also developed to aid the reader in the practical application of these components.

6-1 OUTPUT RECTIFICATION AND FILTERING SCHEMES

The output rectification and filtering scheme used in a power supply depends on the type of supply topology the designer chooses to use. The conventional flyback converter uses the output scheme shown in Fig. 6-1. Since the transformer T_1 in the flyback converter also acts as a storing energy inductor, diode D_1 and capacitor C_1 are the only two elements necessary to produce a dc output. Some practical designs, however, may require the optional

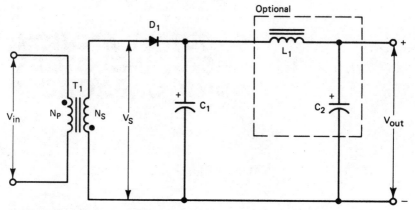

FIGURE 6-1 Output section of a flyback type switching power supply.

insertion of an additional LC filter, shown in Fig. 6-1 within dotted lines, to suppress high-frequency switching spikes. The physical and electrical values of both L and C will be small.

An important factor in the design of any output power supply section is the minimum dc blocking voltage requirement of the rectifier diode and the flywheel diode (for designs which require one). For the flyback converter, the rectifier diode D_1 must have a reverse voltage rating of $[1.2V_{in} (N_S/N_P)]$, minimum.

The output section of a forward converter is shown in Fig. 6-2. Notice the distinct differences in the scheme compared to the flyback. Here, an extra diode D_2, called the flywheel, is added and also inductor L_1 precedes the smoothing capacitor. Diode D_2 provides current to the output during the off period; therefore, the combination of diodes D_1 and D_2 must be capable of delivering full output current, while their reverse blocking voltage capabilities will be equal to $[1.2V_{in} (N_S/N_P)]$, minimum. The output section that is shown in Fig. 6-3 is used for push-pull, half-bridge, and full-bridge converters.

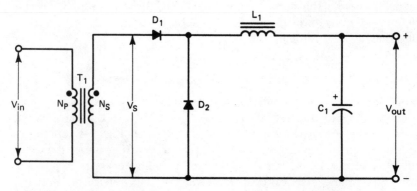

FIGURE 6-2 Output section of a forward type switching power supply.

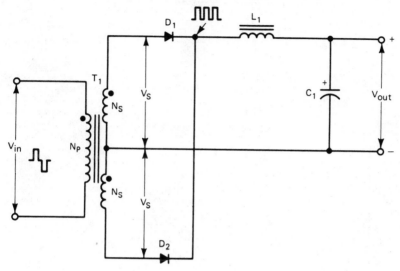

FIGURE 6-3 Output section of push-pull, half-bridge, or full-bridge switching power supply.

Since each of the two diodes D_1 and D_2 provides current to the output for approximately half the cycle, they share the load current equally. An interesting point is that no flywheel diode is needed, because either diode acts as a flywheel when the other one is turned off. Either diode must have a reverse blocking capability of $[2.4V_{out} (V_{in,max}/V_{in,min})]$, minimum.

6-2 POWER RECTIFIER CHARACTERISTICS IN SWITCHING POWER SUPPLY DESIGN

The switching power supply demands that power rectifier diodes must have low forward voltage drop, fast recovery characteristics, and adequate power handling capability. Ordinary PN junction diodes are not suited for switching applications, basically because of their slow recovery and low efficiency. Three types of rectifier diodes are commonly used in switching power supplies: (1) high-efficiency fast recovery, (2) high-efficiency very fast recovery, and (3) Schottky barrier rectifiers. Figure 6-4 shows the typical forward characteristics of these diode types. It can be seen from the graphs that Schottky barrier rectifiers exhibit the smallest forward voltage drop, and therefore they provide higher efficiencies.

The following discussion explores the differences and merits of each rectifier type, and it suggests usage in switch-mode power rectification schemes.

6-2-1 Fast and Very Fast Recovery Diodes

Fast and very fast recovery diodes have moderate to high forward voltage drop, ranging anywhere from 0.8 to 1.2 V. Because of this and because of

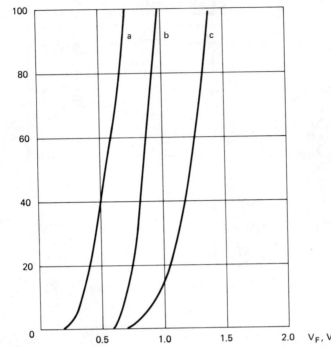

FIGURE 6-4 Typical forward voltage drop characteristics, at various forward current levels, for (*a*) a Schottky barrier rectifier, (*b*) a very fast recovery rectifier, and (*c*) a conventional fast recovery rectifier.

their normally high blocking voltage capabilities, these diodes are particularly suited for low-power, auxiliary voltage rectification for outputs above 12 V.

Because most of today's power supplies operate at 20 kHz and above, the fast and very fast recovery diodes offer reduced reverse recovery time t_{RR} in the nanosecond region for some types. Usually the rule of thumb is to select a fast recovery diode which will have a t_{RR} value at least three times lower than the switching transistor rise time.

These types of diodes also reduce the switching spikes that are associated with the output ripple voltage. Although "soft" recovery diodes tend to be less noisy, their longer t_{RR} and their higher reverse current I_{RM} create much higher switching losses. Figure 6-5 shows the reverse recovery characteristics of abrupt and soft recovery type diodes.

Fast and very fast switching diodes used in switch-mode power supplies as output rectifiers may or may not require heat sinking for their operation, depending upon the maximum working power in the intended application. Normally these diodes have very high junction temperatures, about 175°C,

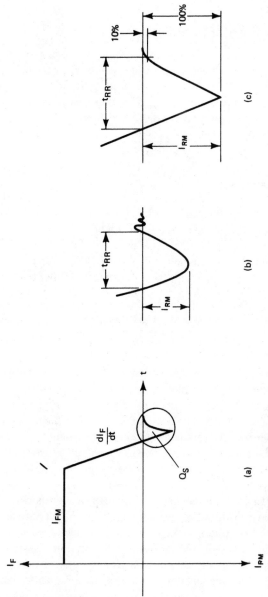

FIGURE 6-5 Waveform (*a*) depicts the behavior of a rectifier as it is switched from forward conduction to reverse at a specific ramp rate *dI/dt*. The circled part of the waveform is the reverse recovery portion. Waveform (*b*) depicts an abrupt recovery type, while (*c*) depicts a soft recovery type. Note the distinct difference in t_{RR} and I_{RM} magnitudes between the two different recovery type diodes.

and most manufacturers give specification graphs, which will allow the designer to calculate the maximum output working current vs. lead or case temperature.

6-2-2 Schottky Barrier Rectifiers

The graph in Fig. 6-4 reveals that the Schottky barrier rectifier has an extremely low forward voltage drop of about 0.5 V, even at high forward currents. This fact makes the Schottky rectifier particularly effective in low voltage outputs, such as 5 V, since in general these outputs deliver high load currents. Moreover, as junction temperature increases in a Schottky, the forward voltage drop becomes even lower.

Reverse recovery time in a Schottky rectifier is negligible, because this device is a majority-carrier semiconductor and therefore there is no minority-carrier storage charge to be removed during switching.

Unfortunately, there are two major drawbacks associated with Schottky barrier rectifiers. First, their reverse blocking capability is low, at present time approximately 100 V. Second, their higher reverse leakage current makes them more susceptible to thermal runaway than other rectifier types. These problems can be averted, however, by providing transient overvoltage protection and by conservative selection of operating junction temperature.

6-2-3 Transient Overvoltage Suppression

Consider the full-wave rectifier, shown in Fig. 6-3, using Schottky rectifiers D_1 and D_2 in a PWM regulated half-bridge power supply. The voltage V_S across each half of the transformer secondary is $2V_{out}$ minimum; therefore each diode must be capable of blocking $2V_S$ at turn-off, or $4V_{out}$.

Unfortunately, the leakage inductance of the high-frequency transformer and the junction capacitance of the Schottky rectifier form a tuned circuit at turn-off, which introduces transient overvoltage ringing, as shown in Fig. 6-6. The amplitude of this ringing may be high enough to exceed the blocking capabilities of the Schottky rectifiers, driving them to destruction during the turn-off period.

The addition of RC snubber networks will suppress this ringing to a safe amplitude, as shown in the lower waveform of Fig. 6-6. There are two ways of incorporating RC snubbers at the output of a power supply to protect the Schottky rectifiers. For high current outputs the snubbers are placed across each rectifier as shown in Fig. 6-7a, while for low current outputs a single RC snubber across the transformer secondary winding, as shown in Fig. 6-7b, may be adequate. Another solution is to place a zener diode, as shown in Fig. 6-7c, to clamp the excessive voltage overshoot to safe levels. Although

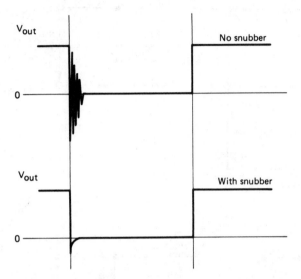

FIGURE 6-6 Upper waveform shows ringing during Schottky rectifier turn-off. Lower waveform shows the same rectifier after transient suppression has been added.

this method works well, the slow recovery of a zener may induce noise spikes at the power supply output, which may not be desirable for low noise applications.

It can be shown that the value of the snubber resistor R_s may be found by equation

$$R_s = \frac{\sqrt{L_T/C_J}}{n} \tag{6-1}$$

where L_T = transformer leakage inductance, μH
$\quad C_J$ = Schottky junction capacitance, pF
$\quad n$ = primary-to-secondary turns ratio, N_P/N_S

The value of the snubber capacitor C_s may be arbitrarily chosen to be anywhere from 0.01 to 0.1 μF.

The power dissipated in the resistor may be found by

$$P_R = \frac{1}{2} C_s \left(\frac{V_{in}}{n}\right)^2 f \tag{6-2}$$

where f is the converter's operating frequency.

Proper selection of the snubber capacitor C_s makes the snubber effective and less dissipative.

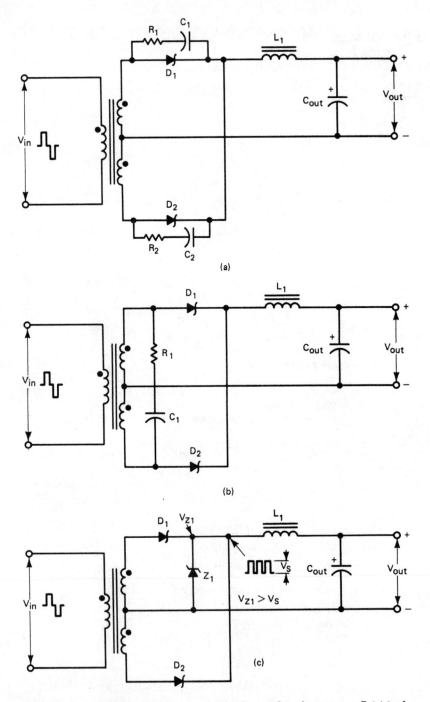

FIGURE 6-7 Means of protecting output Schottky rectifiers during turn-off. (*a*) Snubbers placed across each rectifier; (*b*) a single *RC* snubber placed across the transformer secondary winding; (*c*) using a zener diode.

6-2-4 Calculating the Rectifier Diode Peak Current Capability for the Flyback, Forward, and Push-pull Converters

As mentioned in previous discussions, the output diode in a flyback converter conducts only during part of the converter switching cycle, namely when the switching transistor is off. Accordingly, one has to expect that the output rectifier must have the current capability of providing full output current during the conduction time period (Fig. 6-1).

The minimum peak forward current which the output diode must provide is given by

$$I_{FM} = \frac{2I_{\text{out}}}{1 - \delta_{\text{max}}} \qquad (6\text{-}3)$$

where δ_{max} is the converter's maximum duty factor.

Assuming $\delta_{\text{max}} = 0.45$ for the flyback converter, then

$$I_{FM} = 3.6\ I_{\text{out}} \qquad (6\text{-}4)$$

EXAMPLE 6-1

Calculate the output rectifier peak forward current rating which may be used in a 100-W PWM flyback converter, providing 5 V dc at 20 A output, working with a duty factor $\delta_{\text{max}} = 0.45$, at a frequency of 20 kHz.

SOLUTION

From Eq. 6-4 we get

$$I_{FM} = 3.6\ I_{\text{out}} = 3.6(20) = 72\ \text{A}$$

Therefore a rectifier which can provide 72 A peak repetitive forward current at an approximate duty cycle of 45 percent must be used in this design.

In the forward converter, the selection of output diodes becomes more involved, since one also has to calculate the peak forward current capability of the flywheel diode (Fig. 6-2). On the other hand, because in the forward converter energy is continuously flowing to the output load, the peak forward current will be lower for each diode than in the case of the flyback converter. The peak forward current for the output diodes of a forward converter is given by

$$I_{FM} = I_{\text{out}}\delta_D \qquad (6\text{-}5)$$

where δ_D is the duty factor of either the rectifier or the flywheel diode.

EXAMPLE 6-2

Calculate the rectifier and the flywheel maximum forward current rating for a forward PWM converter with the specifications given in Example 6-1, working over an input voltage range of 90 to 130 V ac.

SOLUTION

The rectifier diode peak forward current is, using Eq. 6-5,

$$I_{FM} = I_{out}\delta_{DR} = 20(0.45) = 9 \text{ A}$$

A rectifier diode with a 10-A rating at 45 percent duty cycle is used. The maximum duty factor of the flywheel diode is

$$\delta_{DF} = 1 - \delta_{min} = 1 - \delta_{max}\left(\frac{V_{in,min}}{V_{in,max}}\right) \qquad (6-6)$$

Take $V_{in,min} = 90\sqrt{2} - 20$ V dc for ripple $= 106$ V dc. We will use 100 V dc. Also, $V_{in,max} = 130\sqrt{2} = 182$. Use 190 V dc. Then from Eq. 6-6

$$\delta_{DF} = 1 - 0.45\left(\frac{100}{190}\right) = 0.76$$

Consequently, the forward current in the flywheel diode is

$$I_{FM} = 20(0.76) = 15.2 \text{ A}$$

A diode rated 20 A at 76 percent duty cycle should be used.

In the push-pull family converters, the output rectifiers provide equal current to the output load during equal conduction cycles (Fig. 6-3). This output scheme is also valid for the half-bridge and the full-bridge circuits.

Because the output of a push-pull converter works as two back-to-back forward converter outputs, maximum forward current for each rectifier is given by Eq. 6-5.

EXAMPLE 6-3

Calculate the maximum forward current rating for each output diode rectifier of a half-bridge PWM converter with the specifications given in Example 6-1.

SOLUTION

The converter's switching period is

$$T = \frac{1}{f} = \frac{1}{20 \text{ kHz}} = 50 \text{ } \mu s$$

Assuming a dead time of 5 μs between each alternative half cycle, then the conduction time for each rectifier is 20 μs. Accordingly, each diode duty factor is $\delta_{DR} = 20/50 = 0.4$. From Eq. 6-5 we get

$$I_{FM} = 20(0.4) = 8 \text{ A}$$

Therefore each rectifier may have a minimum rating of 8 A at 40 percent duty cycle. (In practice 10-A rectifiers should be used.)

As was previously explained, one of the output diode rectifiers acts as a flywheel diode when the other rectifier is turned off. In this case, each diode conducts for 5 μs in the flywheel mode for $\delta_{DF} = 5/50 = 0.1$. Therefore the current provided to the output during the dead time by each rectifier diode is

$$I_{FDM} = 20(0.1) = 2 \text{ A}$$

It is recommended that a thermal analysis be made for the rectifier's intended use, and adequate heat sinking must be provided to avoid destruction due to thermal runaway. Normally a manufacturer provides curves of diode current derating vs. case temperature, which should be consulted during the design.

For higher current outputs, diodes may be paralleled to share the load current. Direct paralleling of diodes must be avoided; instead using separate secondary windings feeding individual diodes, as shown in Fig. 6-8, is recommended.

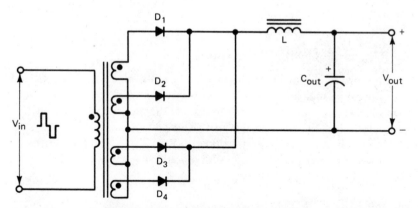

FIGURE 6-8 Separate secondary windings feeding individual diodes should be used where increased output current capacity is needed. Direct diode paralleling must be avoided.

6-3 SYNCHRONOUS RECTIFIERS

6-3-1 General Considerations

As digital integrated circuit manufacturers are working toward implementing more electrical functions and circuits on a single silicon chip, the need for lower bias voltages is necessary. Manufacturers of digital ICs have standardized the new logic bias voltages to 3.3 ± 0.3 V for off-line operation and to 2.8 ± 0.8 V for battery operation.

These new voltage standards have forced design engineers of switching power supplies to explore the possibility of using devices other than junction diodes for rectification, in order to reduce the power losses encountered at the output section of the power supply. For instance, power losses per diode at the output circuit utilizing center-tap rectification topology using conventional rectifiers is given by

$$P = (V_f I_{out}) \delta_{max} \qquad (6\text{-}7)$$

where V_f is the diode forward voltage drop, I_{out} is the power supply rated output current, and δ_{max} is the input waveform duty cycle.

Since the power dissipation is directly proportional to the diode's forward voltage drop, even the best Schottky rectifiers with a 0.4-V forward voltage drop may account for up to 20 percent of the total input power loss. In recent years the power MOSFET with low $R_{DS,on}$ has provided an alternative in power rectification, and as prices of MOSFETs keep dropping, circuits using MOSFETs as synchronous rectifiers have started to appear in switching power supply designs.

Unlike the conventional rectifier the power MOSFET synchronous rectifier exhibits practically no switching losses due to its fast switching times, while its power loss when on and working with a center-tap rectification topology is given by

$$P = I_{out}^2 (R_{DS,on}) \delta_{max} \qquad (6\text{-}8)$$

where $R_{DS,on}$ is the MOSFET on resistance.

Another practical synchronous rectifier is realized using a specialized bipolar transistor, such as the BISYN series offered by Unitrode Corporation. Bipolar synchronous rectifiers match or exceed the merits of their MOSFET counterparts with some added features. The power loss of a bipolar synchronous rectifier is directly proportional to its collector-emitter on resistance and for center-tap applications is given approximately by

$$P = I_{out}^2 (R_{CE,sat}) \delta_{max} \qquad (6\text{-}9)$$

The following paragraphs describe the characteristics of the MOSFET and bipolar synchronous rectifiers in order to familiarize the reader with these devices.

EXAMPLE 6-4

Find the power losses encountered in the output section of a center-tapped full-wave rectifier section of a 3-V, 30-A circuit using the following rectifiers: (a) Schottky diodes with a $V_f = 0.6$ V, (b) power MOSFET synchronous rectifier with $R_{DS,on} = 0.018$ Ω and, (c) BISYN synchronous rectifier with $R_{CE,on} = 0.008$ Ω.

The switching waveform has $\delta_{max} = 0.45$ with a dead-time period of 0.05.

SOLUTION

1. From Eq. 6-7 using Schottky rectifiers: During the positive half cycle when the first diode conducts power, dissipation is

$$P_1 = (V_f I_{out}) \delta_{max} = (0.6)30(0.45) = 8.1 \text{ W}$$

During the dead time t_d, the second diode acts as a flywheel diode dissipating

$$P_2 = (0.6)(3)(0.05) = 0.9 \text{ W}$$

During the negative half cycle of the input waveform, the second diode has the same power dissipation as above. Therefore the total power loss in the circuit is

$$P_t = 2(P_1 + P_2) = 2(8.1 + 0.9) = 18 \text{ W}$$

2. From Eq. 6-8 using the power MOSFET and repeating the above procedure, total power dissipation is found to be

$$P_t = 2(7.29 + 0.81) = 16.2 \text{ W}$$

3. From Eq. 6-9 the total power loss using a BISYN bipolar synchronous rectifier is found to be

$$P_t = 2(3.24 + 0.36) = 7.2 \text{ W}$$

6.3.2 The Power MOSFET as a Synchronous Rectifier

Figure 6-9 shows a practical implementation of a center-tapped full-wave output section of a switching power supply utilizing power MOSFETs as synchronous rectifiers.

MOSFETs Q_1 and Q_2 are selected to have minimum possible $R_{DS,on}$ at maximum output current, in order to increase efficiency. Transformer windings N_3 and N_4 are used to turn on the MOSFETs at opposite half cycles of the input waveform. It is beneficial to overdrive the gate-to-source voltage in order to minimize $R_{DS,on}$. Careful selection of resistors R_1 through R_4 will

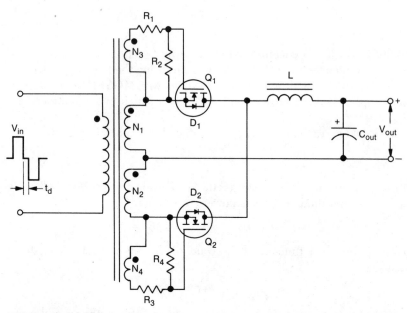

FIGURE 6-9 Full-wave output rectification of a switching power supply using MOS-FETs as synchronous rectifiers. Diodes D_1 and D_2 are parasitic diodes of the MOSFETs.

also minimize the switching times. Diodes D_1 and D_2 are parasitic diodes of the MOSFETs, and they act as free-wheeling diodes during the input waveform dead time t_d to provide current path for inductor L. Since they share the load current during this period, their forward voltage is minimized, but in order for the circuit to be practical, the power loss of each MOSFET when on must be much less than the forward voltage drop of the parasitic diodes, that is,

$$R_{DS,\text{on}}(I_d) \ll I_{FD} \tag{6.10}$$

Figure 6-10 shows the implementation of the MOSFET as synchronous rectifier in a single-ended forward converter. In this configuration transistor Q_1 turns on by the gate-to-source voltage developed by winding N_1, on the positive cycle of the switching waveform. When the input voltage swings negative, MOSFET Q_1 turns off and output current is supplied to the load through MOSFET Q_2, which now turns on through coupled winding N_3 using energy stored in the output inductor L.

The criteria for choosing the proper MOSFET parameters for use in this circuit are the same as in the bridge configuration output circuit. Flyback circuit output sections are implemented as shown in Fig. 6-10, without the use of MOSFET Q_2.

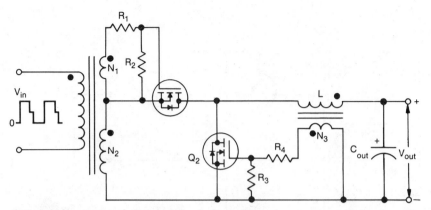

FIGURE 6-10 Circuit implementation of a forward converter output section using power MOSFETs as synchronous rectifiers.

6-3-3 The Bipolar Synchronous Rectifier

Synchronous rectifiers implemented with special bipolar transistors offer all the advantages of the MOSFET synchronous rectifiers up to frequencies of 200 kHz (above this frequency the MOSFET dominates), coupled with lower prices and improved temperature coefficients compared to power MOSFETs.

As an example the BISYN series of bipolar synchronous rectifiers offered by Unitrode Corporation, at their introduction, exhibit very low saturation resistance (8 mΩ for a 4.5-mm chip), with good gain ($\geqslant$25) and symmetrical blocking capabilities for both positive and negative input voltages. The latter feature is very important because now the possibility exists of regulating any output voltage of a multiple-output switching power supply using PWM techniques to drive the bipolar synchronous rectifier, without the need of series-pass linear regulators or magnetic amplifiers, realizing high efficiency gains.

Practical implementation of bipolar synchronous rectifiers is shown in Fig. 6-11 for a center-tapped output circuit. In this circuit synchronous rectifier Q_1 is conducting when the input waveform goes positive providing current to the output. Diode D_3 is free-wheeling, which provides a path for the output inductor current during the input waveform dead-time period t_d. At that instance the voltage across the secondary windings has collapsed, but diode D_1 is forward-biased due to the voltage potential developed across coupled winding N_5 thus speeding up turn-off of synchronous rectifier Q_1.

At the same time diode D_2 current flow will turn on synchronous rectifier Q_2, and it will remain on for the negative duration of the input waveform, until magnetizing current drops below diode D_2 current, i.e., during the positive cycle of the input waveform.

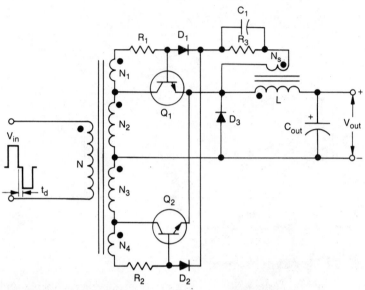

FIGURE 6-11 Bipolar synchronous rectifiers Q_1 and Q_2 used in a center-tapped circuit. (*Courtesy Unitrode Corporation.*)

Figure 6-12 depicts the operation of bipolar synchronous rectifiers in a single-ended forward converter output section.

In the following discussion, it is assumed that the single-ended forward converter has an on-time duty cycle of ≤50 percent. Looking at the circuit when current flows in the primary winding of transformer T_1, voltage is induced in the secondary windings with the polarity shown by the dotted ends. At this time synchronous rectifier Q_1 turns on, allowing current to flow to the output. Because of the polarity of the coupled winding N_4 to the output inductor L, synchronous rectifier Q_2 and diode D_1 are off. During the off period of the input waveform, rectifier Q_2 turns on, sustaining output current through inductor L. Base drive energy is delivered through winding N_4, while diode D_1 still remains off.

When the transformer primary windings are turned on again, the voltage across winding N_3 is clamped to near zero by diode D_1 and the forward-biased base-collector junction of Q_2.

Since winding N_3 is essentially shorted, Q_1 is still off due to lack of base drive, while Q_2 is rapidly shutting off. At that point D_1 is reverse-biased allowing voltage to build up in the secondary windings, turning synchronous rectifier Q_1 on through winding N_1.

The utilization of windings N_3 and N_1 circumvent the long recovery time of the synchronous rectifier Q_2 (300 to 400 ns), thus reducing excess output

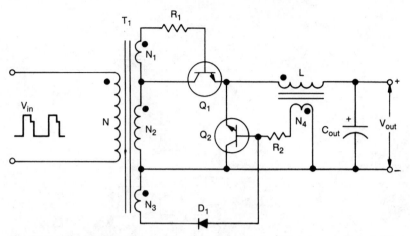

FIGURE 6-12 Implementation of bipolar synchronous rectifiers in a single-ended forward converter output section. (*Courtesy Unitrode Corporation.*)

current overshoot and ringing which, reflected back into the primary, may damage the switching transistor.

6-3-4 Output Voltage Regulation Using Bipolar Synchronous Rectifiers and PWM Regulator Techniques

As we previously mentioned, because of the symmetrical blocking capabilities of the bipolar synchronous rectifier, any output of a switching power supply may be rectified and independently regulated using local PWM techniques.

Figure 6-13 shows a practical implementation of such a circuit, rectifying and regulating a low-voltage, high-current output in a single-ended forward converter, using a Unitrode Corporation BISYN synchronous rectifier and a UC3525A PWM control circuit.

Since the biasing requirements of the BISYN Q_1 necessitate the use of positive and negative power supply voltages, the PWM circuit is biased using ± 12 V generated by rectifying and filtering the secondary voltages developed by windings N_1 and N_3.

Transistor Q_3 is used as a current source to level shift the output voltage in order to drive the inverting input of the PWM error amplifier. The output voltage is developed across resistor R_9 and diode D_4. This diode is used to temperature compensate for the V_{BE} voltage drop of transistor Q_3.

The drive current to the BISYN is limited by resistor R_5 during the PWM on time, except at initial turn-on when the bias drive current may be a few

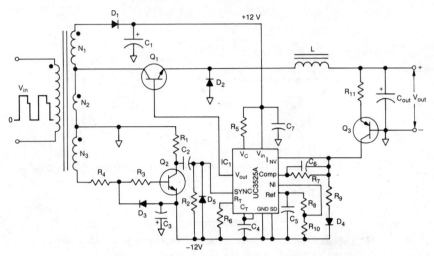

FIGURE 6-13 A BISYN synchronous rectifier Q_1 and a PWM control IC are used to accomplish both rectification and regulation of a 3-V output. (*Courtesy Unitrode Corporation.*)

magnitudes higher than steady-state current. Negative turn-off current is provided by the PWM totem pole output during the off period.

The free running frequency of the PWM is set by resistor R_6 and capacitor C_4 and may be higher than the switching frequency of the power supply itself. Synchronization is accomplished by transistor Q_2, diode D_5, and R_2 and C_2. Diode D_5 clamps the negative voltage excursion of the differentiator circuit R_2/C_2 to prevent malfunction of the control circuit.

6-3-5 A Current-Driven Synchronous Rectifier

A current-driven synchronous rectifier may be implemented using a power MOSFET and a handful of external components, as shown in Fig. 6-14a. The power MOSFET utilized in this design must have a low $R_{DS,on}$ at the required current of operation. During implementation of the circuit shown in Fig. 6-14a, a Motorola TMOS device, the MTM60N05, was used, which has an $R_{DS,on} \leq 30$ mΩ at 10 A.

Referring to Fig. 6-14a, transformer T_1 is a current transformer with windings N_1: N_2: N_3, having a $1:25:3$ turns ratio. A voltage applied at terminal A will cause a source current I_s to flow through winding N_1. In turn, a current of $0.04I_s$ A is induced in winding N_2, flowing through diode D_1 into the gate of power MOSFET Q_1. This current will charge the FET input capacitance C_{gs}, turning the transistor on. Transformer T_1 is designed so that it will saturate at some point. Saturation of the core terminates the charging of the FET input capacitance, since all transformer windings appear shorted. Con-

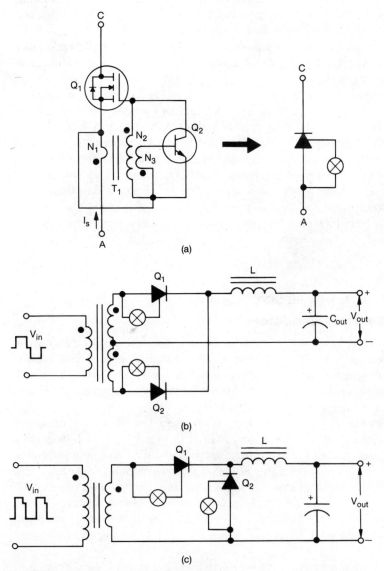

FIGURE 6-14 (*a*) A current-driven synchronous rectifier circuit and its suggested symbol. (*b*) and (*c*) Applications of the synchronous rectifier in a full-wave center-tapped rectification scheme and a single-ended forward converter output scheme. (*Courtesy Motorola Semiconductor Products, Inc.*)

151

sequently, because of the polarity of winding N_3, transistor Q_2 remains off throughout the charging and terminating cycle, diode D_1 is reverse-biased, and the input capacitance holds its charge. In this state, Q_1 is fully enhanced allowing all the I_s current to flow through it. Due to the low $R_{Ds,on}$ of the FET, Q_1 is acting as a high-efficiency rectifier diode. The stored magnetizing current in the transformer T_1, if unopposed, will cause the windings to reverse their polarity.

When this happens the base of bipolar transistor Q_2 becomes positive, turning it on, thus discharging the FET input capacitance turning Q_1 off, which in turn blocks the flow of I_s. Resetting of the transformer core takes place at this time also. The base-emitter drop of Q_2 clamps the reset and guarantees that Q_2 is conducting during the critical high dv/dt recovery interval of Q_1. This prevents false turn-on of Q_1 under any circumstance.

The absence of any external control or timing signals makes this circuit a true two-terminal rectifier. Figures 6-14b and c show how this rectifier circuit may be implemented in a full-wave center-tapped rectification scheme or in a single-ended scheme.

6-4 OUTPUT POWER INDUCTOR DESIGN

6-4-1 General Considerations

Most switching power supply designs use an inductor as part of their output filtering configuration. The presence of this inductor is two-fold: first it stores energy during the off or "notch" periods in order to keep the output current flowing continuously to the load, and second it aids to smooth out and average the output voltage ripple to acceptable levels.

There are a variety of cores that an engineer can use in the design of inductors. the most popular materials used in present-day high-frequency switching designs are ferrite cores, iron powder cores, and molypermalloy (MPP) cores. All of these cores are good for power inductor designs, and basically the criterion of choosing one vs. the other is based on factors such as cost, weight, availability, performance, and ease of manufacture.

Iron powder and MPP cores are generally offered in toroid forms, and they are well suited for power chokes because of the following characteristics:

1. High saturation flux density B_{sat} up to 8000 G.

2. High energy storage capability.

3. Inherent air gap eliminates the need of gapping the core.

4. Wide choice of sizes.

Ferrite cores, on the other hand, have to be gapped because of their low saturation flux density B_{sat}, they are more temperature sensitive, and they tend to be bulkier. But if pot cores are used for output chokes, radiated EMI will be reduced because of the inherent shielding properties of the pot core. Also, ferrite chokes are easier to wind, especially if heavy-gauge wire is involved.

6-4-2 Deriving the Design Equations

Consider the output section of a PWM half-bridge converter depicted in Fig. 6-15a. The output waveforms E_{in} and E_{out} are shown in Fig. 6-15b, as well as the average load current I_{out} with ripple ΔI riding on it.

From basic electrical theory, the voltage across the inductor is given by

$$V_L = L \frac{di}{dt} \tag{6-11}$$

Since

$$V_L = E_{in} - E_{out}$$

and

$$di = \Delta I_L$$

then Eq. 6-7 may be written, solving for L, as follows:

$$L = \frac{(E_{in} - E_{out})\Delta t}{\delta I_L} \tag{6-12}$$

In the case of the PWM half-bridge or full-bridge converter, the voltage E_{in} is roughly twice the value of the output voltage E_{out} at the maximum primary input voltage V_{in} (see Fig. 6-15). Therefore, $E_{in} - E_{out} = E_{out}$. The time interval Δt is equal to the maximum dead time, or "notch" time, t_{off}, which occurs between alternate switching half cycles.

Maximum t_{off} occurs at maximum input line voltage, since the transistor conduction time t_{on} is at a minimum. Therefore, the inductor must be designed to store enough energy to provide continuous output current during the notch periods.

Expressing Δt in terms of secondary voltage E_{in} and E_{out} yields

$$t = t_{off} = \frac{1}{2}\left[\frac{1 - (E_{out}/E_{in})}{f}\right] \tag{6-13}$$

where f is the converter's frequency in kilohertz, while the factor 1/2 relates the notch time t_{off} to the entire switching cycle, since the total switching

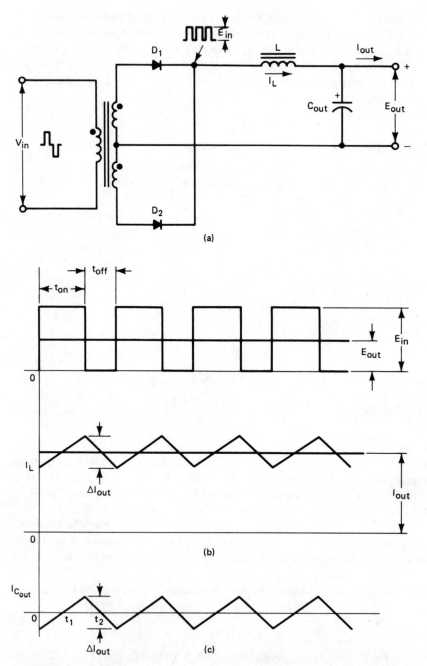

FIGURE 6.15 (*a*) The output section of a PWM half-bridge converter, and (*b, c*) its associated voltage and current waveforms.

period encounters two notch time intervals. In order to keep low inductor peak current and good output ripple, it is recommended that ΔI_L not be greater than $0.25 I_{out}$.

Based on the above, Eq. 6-7 may be rewritten as follows:

$$L = \frac{E_{out} t_{off}}{0.25 I_{out}} \tag{6-14}$$

Eq. 6-10 gives an inductance value which will be very close to the practical value, and it may or may not require fine tuning in the actual application. After the inductance has been calculated, the core size and core material has to be chosen in order to complete the design.

The following examples show step-by-step design procedures using a ferrite core and an MPP core. The first design procedure is analytical, while the second is graphical. Both methods are equally useful for designing optimum filter chokes.

EXAMPLE 6-5

For a 20-kHz, 100-W half-bridge power supply having an output of 5 V dc at 20 A, calculate the output inductor L using a ferrite core.

SOLUTION

Using Eq. 6-13 to calculate the maximum notch period,

$$t_{off} = \frac{1 - (E_{out}/E_{in})}{2f} = \frac{1 - (5/10)}{40 \times 10^3} = 12 \ \mu s$$

The inductor value L capable of delivering output current ΔI_L during t_{off} is

$$L = \frac{E_{out} t_{off}}{0.25 I_{out}} = \frac{5 \times 12 \text{ s}}{5} = 12 \ \mu H$$

Select the minimum size of core using the following equation

$$A_e A_c = \frac{(5.067) \ 10^8 \ (L I_{out} D^2)}{K B_{max}}$$

where K = 0.4 for toroids and 0.8 for bobbins
$\quad \ \ D$ = diameter of wire to be used
$\quad \ \ A_e$ = core effective area
$\quad \ \ A_c$ = bobbin winding area

Choose a current density of 400 c.m./A. Then for 20 A, the wire is 400 × 20 = 8000 c.m., which corresponds to no. 11 AWG wire, with a maximum diameter of 0.0948 from Table 5-2.

Also selecting a B_{max} = 2000 G, the $A_e A_c$ product is

$$A_e A_c = \frac{5.067 \times 10^8 \times 12 \times 10^{-6} \times 20 \times 0.0948^2}{0.8 \times 2000} = 0.683 \text{ cm}^4$$

From ferrite catalogs the 3019 pot core has A_e = 1.38 cm^2 and A_c = 0.587 cm^2, which yields $A_e A_e$ = 0.81 cm^4. This is good enough for our requirements, but we may have to use a larger core in order to fit the large-gauge wire. In fact, it is advisable to use a bundle of smaller wire size to increase the conductor surface area and reduce skin-effect losses. For this example, eight no. 20 AWG wires will have to be used in place of one no. 11 AWG wire. By using bundled conductors in parallel, copper losses due to reduced I^2R losses are minimized; therefore we could reduce the current density requirement. Using only six no. 20 magnet wires, the current density will be 300 c.m./A, still an acceptable value. With all of this in mind, the 3622 core and single-section bobbin is chosen for this design.

Because the inductor experiences a large dc bias, it is necessary to gap the core in order to avoid saturation. The length of the gap is

$$l_g = \frac{(0.4\pi L I_{out}^2)10^8}{A_e B_{max}^2} = \frac{0.4 \times 3.14 \times 12 \times 10^{-6} \times 20^2 \times 10^8}{2.02 \times 2000^2}$$

$$= 0.0746 \text{ cm}$$

Since the air gap interrupts the magnetic circuit twice, if a spacer is used to provide the gapping, the spacer thickness will be $l_g/2$ = 0.0373. On the other hand, the total gap length must be used if only the center leg is gapped.

Now the number of turns may be calculated.

$$N = \frac{B_{max} l_g}{0.4\pi I_{out}} = \frac{2000 \times 0.0746}{0.4 \times 3.14 \times 20} = 5.94 \text{ turns}$$

We will use six turns. Using six no. 20 wire conductors in parallel, the equivalent of 6 × 6 = 36 turns would be required.

The turns per bobbin graph (refer to a Ferroxcube catalog) on a 3622 single-section bobbin data sheet shows that approximately 60 turns of no. 20 wire will fill the bobbin winding area. Taking into account air space around the round wires and tape finish, we conclude that the 3622 bobbin and pot core is a good choice for this design. Improvements may be made in the actual application, which could include increasing the number of turns for better filtering or increasing the number of conductors to reduce heating effects.

EXAMPLE 6-6

Design the filter choke of Example 6-5 using an MPP core.

SOLUTION

Although the design of a filter choke using MPP cores may be analytically accomplished, this example presents a quick graphical method, developed by Magnetics, Inc., which is fast and accurate. The example is based on Magnetics, Inc., MPP core data, but many manufacturers make equivalent core sizes with similar magnetic properties; thus the reader may extend the described method to the manufacturer of choice.

Step 1: Calculate the required inductance. Using Eq. 6-14,

$$L = \frac{E_{out}t_{off}}{0.25I_{out}} = \frac{5 \times 12}{5} = 12 \ \mu H$$

Step 2: Calculate the product of LI_{out}^2. Taking $L = 12 \ \mu H$ and $I_{out} = 20$ A, then $LI^2 = (12 \ \mu H)(20^2) = (0.012 \ mH)(20^2) = 4.8$.

Step 3: Select core size. From the core selector chart of Fig. 6-16, locate the LI^2 point of 4.8; following this coordinate the first core size encountered that falls within the solid line permeability family is size 55548.

Step 4: Select the permeability. The intersection point of the 4.8 coordinate and the 55548 core size coordinate falls between the 26 μ and 60 μ curves. Only those permeability lines which intersect the LI^2 coordinate (4.8 in this case) below the core intersection may be used. In this case we will use a core with a permeability of 60 as a first try. If a higher permeability core (such as $\mu = 125$) is used, this choice will yield a lower winding factor, therefore fewer turns.

Step 5: Calculate the number of turns to obtain the required inductance. The number of turns can be calculated as follows:

$$N = 1000\sqrt{\frac{L}{L_{1000}}} \tag{6-15}$$

where L is the desired inductance (in millihenries), and L_{1000} is the nominal inductance (in millihenries per 1000 turns).

From Table 6-1 or Table 6-2, we find $L_{1000} = 61$ for the 55548 core with a permeability of 60 μ. Therefore the required number of

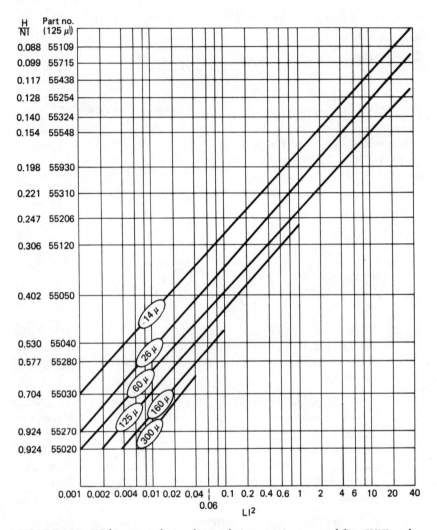

FIGURE 6-16 DC bias core selector chart, relating core size, permeability, *H/NI*, and *LI²* for MPP cores. *L* = inductance with dc bias (in millihenries); *I* = dc current (in amperes). (*Courtesy of Magnetics, Inc.*)

turns to obtain an inductance of 12 μH (0.012 mH) is

$$N = 1000 \sqrt{\frac{0.012}{61}} = 14 \text{ turns}$$

Increasing the number of turns by 20 percent we get $N = 17$ turns.

TABLE 6-1 ELECTRICAL, MECHANICAL, AND WINDING INFORMATION DATA ON MAGNETICS, INC., 55548 MPP FAMILY CORES

CORE DIMENSIONS AFTER FINISH		
DD (Max.)	1.332 in.	33.80 mm
ID (Min.)	0.760 in.	19.30 mm
HT (Max.)	0.457 in.	11.61 mm

1.300
0.785
0.420

Window area | 577,600 c.m.
Cross section | 0.1042 in.^{2} | 0.672 cm^2
Path length | 3.21 in. | 8.15 cm
Weight | 1.7 oz | 47. gm

WINDING TURN LENGTH

Winding factor		Length/turn
100% (Unity)	0.1943 ft	5.93 cm
60%	0.1668 ft	5.09 cm
40%	0.1400 ft	4.27 cm
20%	0.1282 ft	3.91 cm
0%	0.1238 ft	3.78 cm

WOUND COIL DIMENSIONS
Unity winding factor

DD (Max.)	1.840 in.	46.7 mm
HT (Max.)	1.103 in.	28.0 mm

MAGNETIC INFORMATION

Part no.	Perm., μ	Inductance @ 1000 turns, MH ± 8%	Nominal dc resistance, Ω/MH	Finishes and stabiliza-tions	Grading status, 2% bands	B/NI Gauss per amp. turn	
55551 –	14	14	0.335	A2	*	2.16	(<1500 G)
55550 –	28	28	0.167	A2	*	4.00	(<1500 G)
55071 –	60	61	0.0768	ALL	Yes	9.24	(<1500 G)
55548 –	125	127	0.0369	ALL	Yes	19.3	(<1500 G)
55547 –	147	150	0.0312	ALL	Yes	22.6	(<1500 G)
55546 –	160	163	0.0287	ALL	Yes	24.6	(<1500 G)
55542 –	173	176	0.0266	ALL	Yes	26.6	(<1500 G)
55545 –	200	203	0.0230	ALL	Yes	30.8	(<600 G)
55543 –	300	305	0.0153	A2 and L8	Yes	46.2	(<3500 G)
55544 –	550	559	0.0083	A2	Yes	84.7	(<50 G)

WINDING INFORMATION
for unity winding factor

AWG wire size	Turns	Rdc Ω	AWG wire size	Turns	Rds Ω
8	32	0.00393	23	889	3.50
9	40	0.00618	24	1100	5.49
10	50	0.00976	25	1359	8.56
11	63	0.01544	26	1699	13.53
12	79	0.0244	27	2139	21.4
13	99	0.0384	28	2625	33.3
14	123	0.0604	29	3209	51.3
15	154	0.0949	30	4011	81.1
16	193	0.1504	31	4937	125.7
17	239	0.234	32	6017	189.4
18	298	0.370	33	7463	299
19	370	0.579	34	9500	482
20	462	0.909	35	11,788	758
21	578	1.437	36	14,549	1173
22	713	2.24			

Source: Courtesy of Magnetics, Inc.

TABLE 6-2 INDUCTANCE TABLE

Part no., 125 μ	Inductance per 1000 turns, mH									
	14 μ	26 μ	60 μ	125 μ	147 μ	160 μ	173 μ	200 μ	300 μ	550 μ
55140	NA	NA	NA	26	31	33	36	42	62	NA
55150	4	7	17	35	41	45	48	56	84	NA
55180	5	9	20	42	49	53	57	67	99	NA
55020	6	10	24	50	59	64	69	80	120	220
55240	6	11	26	54	64	69	75	86	130	242
55270	12	21	50	103	122	132	144	165	247	466
55030	6	11	25	52	62	66	73	83	124	229
55280	6	11	25	53	63	68	74	84	128	232
55290	7	14	32	66	78	84	92	105	159	290
55040	7	14	32	66	78	84	92	105	159	290
55130	6	11	26	53	63	68	74	85	127	NA
55050	6.4	12	27	56	67	72	79	90	134	255
55120	8	15	35	72	88	92	104	115	173	317
55206	7.8	14	32	68	81	87	96	109	163	320
55310	9.9	19	43	90	106	115	124	144	216	396
55350	12	22	51	105	124	135	146	169	253	NA
55930	18	32	75	157	185	201	217	251	377	740
55548	14	28	61	127	150	163	176	203	305	559
55585	9	16	38	79	93	101	109	126	190	348
55324	13	24	56	117	138	150	162	187	281	515
55254	19	35	81	168	198	215	233	269	403	740
55438	32	59	135	281	330	360	390	450	674	NA
55089	20	37	86	178	210	228	246	285	427	NA
55715	17	32	73	152	179	195	210	243	365	NA
55109	18	33	75	156	185	200	218	250	374	NA
55866	16	30	68	142	NA	NA	NA	NA	NA	NA

Note: Relates Magnetics, Inc., MPP core number to inductance per 1000 turns at various permeability values.
Source: Courtesy of Magnetics, Inc.

Step 6: Calculate wire size and fit. If we choose a current density of 400 c.m./A, then a 400 c.m./A × 20 A = 8000-c.m. wire is needed. This corresponds to no. 11 AWG wire from Table 5-2.

To reduce skin effect losses, four no. 17 AWG wires in parallel will be used, the equivalent of 17 × 4 = 68 turns of single no. 17 wire. To check for fit, 68 turns of no. 17 wire (2050 c.m.) equals 139,400 c.m. From Table 6-1, the 55548 family of MPP cores has a total window area of 577,600 c.m. Therefore, the winding factor of this core is equal to 139,400/577,600 = 0.24. Also, from Fig. 6-15, the winding information data show that a fully wound core will accept 239 turns of no. 17 AWG wire. For a winding factor of 24 percent the number of turns which will fill the core will be 239 × 0.24 = 57.36 turns. Since our design requires 68 turns, a 55548 with higher permeability should have been chosen.

Choosing the next higher permeability of 125 μ from Table 6-2, $L_{1000} = 127$. Thus, step 5 is corrected to give the necessary number of turns to achieve the required inductance.

$$N = 1000 \sqrt{\frac{0.012}{127}} = 9.72 \text{ turns}$$

Increasing the number of turns by 20 percent, $N = 12$ turns of no. 11 AWG wire is required, or 12 × 4 = 48 turns of four no. 17 AWG wires in parallel, which will fit the chosen core.

In order to check the results, the following analysis may be used.

Step 1: Calculate the dc magnetizing force. From Table 6-1 the 55548 core shows H/NI = 0.154. Then, the magnetizing force will be

$$H = \frac{H}{NI} (NI) = 0.154 \times 12 \times 20 = 36.98 \text{ Oe}$$

Step 2: Check permeability reduction. From the curves of Fig. 6-17, the 125-μ material at 36.98 Oe experiences a reduction of 30 percent of its initial permeability; therefore, the percent of useful permeability is 70 percent.

Step 3: Find the inductance of the core at the permeability of step 2. The nominal inducatnce of the 55548 core for 125-μ material is 127 mH

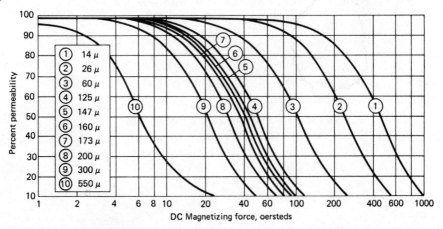

FIGURE 6-17 Permeability vs. dc bias curves for Magnetics, Inc., MPP cores. (*Courtesy of Magnetics, Inc.*)

per 1000 turns. At 70 percent permeability the nominal inductance becomes $127 \times 0.70 = 88.5$ mH per 1000 turns. Solving for L, Eq. 6-15 yields

$$L = \left(\frac{N}{1000}\right)^2 (L_{1000}) = \left(\frac{12}{1000}\right)^2 (88.5) = 12.74 \; \mu\text{H}$$

Therefore the minimum inductance of 12 μH has been achieved.

6-5 MAGNETIC AMPLIFIERS

The magnetic-amplifier technology has been around for a long time, but renewed interest has emerged lately, especially with the proliferation of the switching regulator, as an efficient means of regulating the auxiliary outputs of a multiple-output switching power supply.

The term "magnetic amplifier" is really misleading, since the circuit neither amplifies nor does it use any amplifier as its main switching element. Rather, an inductive element is used as a control switch. Thus a magnetic amplifier is a reactor, wound on a core with relatively square B-H characteristics. This reactor has two distinct modes of operation: When unsaturated, it acts as an inductance capable of supporting a large voltage with very little or no current flow, and when saturated, the reactor impedance drops to zero, allowing current to flow with zero voltage drop.

The magnetic amplifier in itself is a pulse-width-modulated buck regulator and requires an output LC filter to convert its PWM output to a dc voltage. Since it is a buck regulator, it can only lower the output voltage from what it would be with the regulator bypassed.

Magnetic-amplifier usage is universal, that is, they work equally well with any converter topology such as forward, flyback, and push-pull and their derivatives.

6-5-1 Operation of the Magnetic Amplifier

Figure 6-18 shows a simplified magnetic amplifier and its corresponding waveforms. Let us examine the operation of the circuit.

Assume that N_s is the secondary winding of a transformer driven from a square wave which produces a ± 12-V waveform at V_1. At time $t = 0^-$, the reactor L_C is saturated and $V_2 = V_3 = 12$ V (diode drops in this discussion are ignored for the sake of simplicity, but they should be considered in an actual design). At time $t = 0^+$ to $t = 10 \; \mu$s, the reactor has a voltage across it of $V_{L_C} = -12 - (-6) = -6$ V, since $V_C = -6$ V. Thus a reset current flows from V_C through D_1 to the reactor for a period of 10 μs, driving the core out of saturation and resetting it by an amount equal to 60 V-μs.

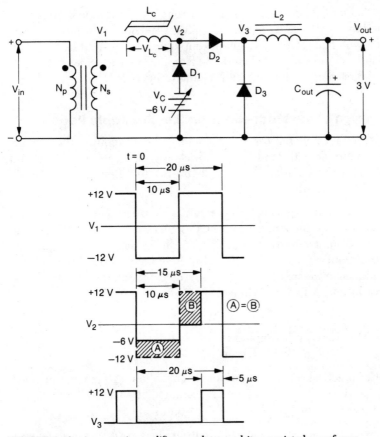

FIGURE 6-18   A magnetic-amplifier regulator and its associated waveforms.

At time $t = 10^+$ μs, V_C switches positive again, and the core should be driven into saturation. But since the core has to be reset by a volt-seconds product equal to the amount which took it out of saturation, that is, 60 V-μs, the mag amp delays the leading edge of the waveform by 5 μs. This condition remains until the voltage across the core drives the reactor into saturation, delivering a 5-μs-long output pulse. In mathematical terms we have the following relationships:

$$A = 6 \text{ V} \cdot 10 \text{ }\mu s = 60 \text{ V-}\mu s$$
$$B = 12 \text{ V} \cdot 5 \text{ }\mu s = 60 \text{ V-}\mu s$$

Therefore the output voltage will be

$$V_{out} = V_{in}\left(\frac{t_{on}}{T}\right) = 12\left(\frac{5}{20}\right) = 3 \text{ V}$$

It is important to note that the reset current of a magnetic amplifier is determined by the core and the number of turns and not by the load current. Hence, a few milliamperes of magnetic-amplifier reset current can and will control many amperes of load current. In fact, magnetic amplifiers are cost-effective and efficient at outputs with load currents over 2 A.

6-5-2 Design of the Magnetic-Amplifier Saturable Reactor

Designing the saturable reactor of a magnetic amplifier requires three steps: First, determine the withstand Λ; second, choose the appropriate core; and third, calculate the number of turns of the reactor. Let us look at each individual step separately.

Step 1: Assuming that the output inductor has been designed for continuous conduction, the reactor must be designed to delay the leading edge of the input waveform to the filter inductor long enough to provide the required output voltage. This excluded portion of the pulse is equivalent to area B, as shown in Fig. 6-18, and it is defined as the withstand Λ in V-μs. Hence,

$$\Lambda = Vt \qquad (6\text{-}16)$$

where V = pulse amplitude, V
  t = leading-edge delay, μs

For practical considerations one should allow a 20 percent increase over the required withstand to accommodate an equivalent increase or decrease in anticipated load current changes.

Step 2: Select the wire size based on output current. A practical value of 500 c.m./A (circular mils per ampere) is an acceptable design rule. Next choose the core material to determine the saturation flux density B_{max}. For best results choose a core material with an outside diameter of 1 mil or less. Table 6-3 provides several commonly used materials, and it is presented here as a guide to the reader. Choose a fill factor K using values from 0.1 to 0.3, with the lower values for large wire sizes. Next calculate the core size based upon the area product:

$$W_a A_c = \frac{A_w \Lambda 10^8}{2(B_{max})K} \qquad (6\text{-}17)$$

where W_a = core window area, cm^2
  A_c = effective core area, cm^2
  A_w = wire area, cm^2
  Λ = required withstand, V-s
  B_{max} = core saturation flux density, G
  K = fill factor

TABLE 6-3 MATERIALS COMPARISON

Material type	Flux density (Kilogausses)	Squareness	Coercive force d.c.	Coercive force 400 cps CCFR**	Gain***
Magnesil	15.0–18.0	0.85 up	0.4–0.6	0.45–0.65	130–220
Square Orthonol	14.2–15.8	0.94 up	0.1–0.2	0.15–0.25	310–715
48 Alloy	11.5–14.0	0.80--0.92	0.05–0.15	0.08–0.15	280–550
Square Permalloy 80	6.6–8.2	0.80 up	0.02–0.04	0.022–0.044	550–1650
Round Permalloy 80	6.6–8.2	0.45–0.75	0.008–0.02	0.008–0.026	250–715
Supermalloy	6.5–8.2	0.40–0.70	0.003–0.008	0.004–0.015	250–715
Supermendur	19–22	0.90 up	0.15–0.35	0.50–0.070	85–135
Metglas 2605SC	15–16	0.90 up	0.03–0.08	0.04–0.1	400–900
Metglas 2714A	5–6.5	0.90 up	0.008–0.02	0.01–0.025	750–2500

*The values listed are typical of .002″ thick materials (Metglas .001″) of the types shown. For guaranteed characteristics on all thicknesses of all alloys available, refer to Magnetics Inc.'s Guaranteed Tape Wound Core Characteristics Bulletin, which is available upon request, from the Components Sales Department, Magnetics Inc., Butler, Pa.

**400 cycle CCFR Coercive Force is defined as the H^1 reset characteristics described by the Constant Current Flux Reset Test Method in AIEE Paper #432.

***Gain is the 400 cycle core. Gain described by the Constant Current Flux Reset Test Method per AIEE Paper #432 for cores with ID/OD of .75 to .80.

Source: Courtesy of Hagnetics, Inc.

Step 3: Determine the appropriate number of turns using the formula

$$N = \frac{\Lambda(10^8)}{2(B_{max})A_c} \qquad (6\text{-}18)$$

and estimate the control current using the formula

$$I_c = \frac{(0.796)Hl_e}{N} \qquad (6\text{-}19)$$

where H = Magnetizing force, Oe
l_e = Magnetic path length, cm

To better understand magnetic-amplifier operation, the following example gives the reader all the necessary steps to design efficient mag amps.

EXAMPLE 6-7

Consider the forward converter given in Fig. 6-19 and its associated waveforms. Design the magnetic amplifier to produce a regulated 12-V, auxiliary output voltage at 8 A. Switching frequency is 100 kHz.

SOLUTION

Assume that the pulse height at V_1 is 40 V, at 100 kHz switching frequency. To produce 12 V at the output, the average value of the voltage at V_2 must be 12 V, and the required positive pulse width will be

$$PW = \frac{V_2}{(V_1)f} = \frac{12}{(40)(100 \times 10^3)} = 3 \; \mu s$$

Assuming a "dead time" at the switching waveform of 2 μs gives a 4-μs input pulse width, as shown in Fig. 6-19, for a total period of 10 μs (100 kHz).

With a 4-μs-wide pulse at V_1, the saturable core L_c must delay the leading edge by 1 μs, to achieve the required 3-μs pulse width. The core then must withstand

$$\Lambda = Vt = (40)(1) = 40 \; V\text{-}\mu s$$

During reset of the core, the withstand must be equal during the negative half cycle. The value of the reverse voltage required to achieve equal withstand integrals is

$$V_R = \frac{40}{4} = 10 \; V$$

Hence, the negative half cycle of the waveform will be clamped by diode D_1 at $-40 - (-10) = -30$ V, as depicted by the V_2 waveform of Fig. 6-19.

Step 1: Having calculated the withstand Λ to be 40 V-μs, we allow a 20 percent tolerance for a total maximum withstand of 48 V-μs.

Step 2: To calculate the required wire size to be used in the saturable reactor, the root-mean-square (rms) value of the output current is calculated. Thus with a duty cycle of $\delta = 12/40 = 0.3$, the root-mean-square current is

$$I_{rms} = \sqrt{I^2 \delta} = \sqrt{(8)^2(0.3)} = 4.4 \; A$$

Using 500 c.m./A, the wire area is calculated to be 500 c.m./A $\times$ 4.4 A = 2200 c.m. From Table 5-2, the closest wire gauge corresponding to this value is 16.

For this example we choose as the core material a Square Permalloy 80, which has a very low coercive force and a very square B-H loop, with a $B_{max} = 7000$ G (see Table 6-3). Thus we can now calculate $W_a A_c$ from Eq.

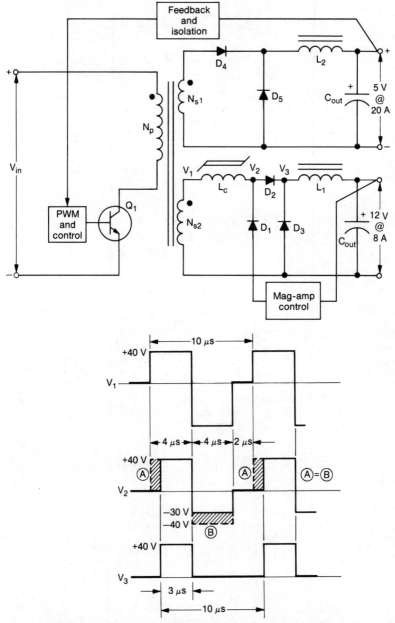

FIGURE 6-19 A 100-kHz forward converter using a magnetic amplifier and its associated waveforms, for Example 6-7.

TABLE 6-4 HIGH-FREQUENCY MAG-AMP CORES

Part number	I.D. (in.) Core	I.D. Case (min)	O.D. (in.) Core	O.D. Case (max.)	Ht. (in.) Core	Ht. Case (max.)	Core loss (w) @ 50 kHz, 2000 gauss (max.)	l_e cm	A_c cm²	W_a cr m	Core wt. grams	$W_a A_c$ cr m cm² ($\times 10^{-6}$)
50B10-5D	0.650	0.580	0.900	0.970	0.125	0.200	0.118	6.18	0.051	308,000	2.7	0.0157
50B10-1D	0.650	0.580	0.900	0.970	0.125	0.200	0.22	6.18	0.076	308,000	4.0	0.0234
50B10-1E	0.650	0.580	0.900	0.970	0.125	0.200	0.092	6.18	0.076	308,000	3.5	0.0234
50B11-5D	0.500	0.430	0.625	0.695	0.125	0.200	0.044	4.49	0.025	142,000	1.0	0.0035
50B11-D	0.500	0.430	0.625	0.695	0.125	0.200	0.083	4.49	0.038	142,000	1.5	0.0054
50B11-1E	0.500	0.430	0.625	0.695	0.125	0.200	0.034	4.49	0.038	142,000	1.3	0.0054
50B12-5D	0.375	0.305	0.500	0.570	0.125	0.200	0.035	3.49	0.025	99,000	0.8	0.0025
50B12-1D	0.375	0.305	0.500	0.570	0.125	0.200	0.066	3.49	0.038	99,000	1.2	0.0038
50B12-1E	0.375	0.305	0.500	0.570	0.125	0.200	0.027	3.49	0.038	99,000	1.04	0.0038

Note: These cores are specifically designed for this application. (5D = ½ mil permalloy, 1D = 1 mil permalloy, 1E = 1 mil METGLAS ALLOY 2714A.) Contact the factory for other sizes.

Source: Courtesy of Magnetics, Inc.

6-17, as follows:

$$W_a A_c = \frac{(2200)(48)(10^{-6})(10^8)}{2(7000)(0.1)} = 75.4 \times 10^2 \text{ c.m. cm}^2$$

$$= 0.00754 \times 10^6 \text{ c.m. cm}^2$$

Note that since the wire size is 16 gauge, which is relatively large, a fill factor of 0.1 has been chosen as a sensible compromise.

Since the $W_a A_c$ calculated above must be at least 0.00754×10^6 c.m. cm^2, and because the converter frequency is 100 kHz, a core with a tape thickness of 0.0005 in. ($\frac{1}{2}$ mil) will be used.

Table 6-4 lists a choice of high-frequency mag-amp cores. From the table, the logical candidate for this application is the 50B10-5D core. Knowing the type of core, the number of turns required may be easily calculated using

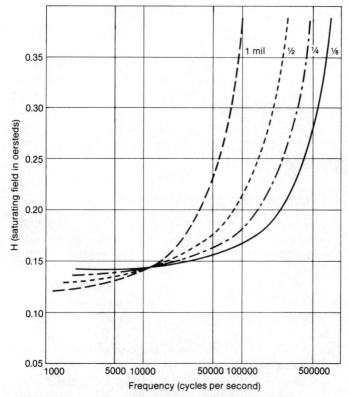

FIGURE 6-20 Average MMF required to saturate Permalloy 80 vs. frequency using square-wave current drive. (*Courtesy Magnetics, Inc.*)

Eq. 6-18. Hence

$$N = \frac{\Lambda(10^8)}{2(B_{max})A_c} = \frac{48 \times 10^{-6} \times 10^8}{2(7000)(0.051)} = 6.7 \text{ turns}$$

Round off to 7 turns.

Completing the example, the magnetizing current is calculated, in order to estimate the current required to reset the core when designing the control circuit.

From Fig. 6-20 the magnetizing force at 100 kHz for a $\frac{1}{2}$-mil core is $H = 0.215$ Oe. Thus the control current will be

$$I_c = \frac{(0.796)Hl_e}{N} = \frac{(0.796)(0.215)(6.18)}{7} = 0.15 \text{ A}$$

Note that the circuit in this example was designed for regulation only. When independent shutdown is added for short-circuit protection or turn-off from an external logic signal, then the required withstand is the area under the entire positive input pulse. In this example, $\Lambda = (40 \text{ V})(4 \text{ s}) = 160 \text{ V-}\mu\text{s}$, and all the above calculations must be repeated based upon this withstand value.

As mentioned before, the magnetic amplifier is a buck regulator, that is, its output voltage will always be lower than its input voltage. Since the power losses of the magnetic amplifiers are small, these circuits are proving to be ideal candidates for deriving low-voltage, high-current outputs, that is, 3 V or lower, at relatively high efficiencies. The applications of magnetic amplifiers to switching power supply design are countless and may be limited only by the creativity and imagination of the desigh engineer.

6-5-3 Control Circuits for Magnetic Amplifiers

Figure 6-21 shows a fundamental magnetic-amplifier control circuit regulating the output of a forward converter. In this circuit, reset of the saturable core is accomplished by transistor Q_1 and its associated components. Thus, during the negative half cycle of the source voltage, transistor Q_1 conducts allowing a current to flow through R_1 and D_3, resetting the saturable core L_c. The reset flux level is controlled by the transistor equivalent resistance, which is a function of the difference between the output and reference voltages.

The drawback of this circuit is its temperature sensitivity and its tendency to oscillate at extreme load changes. An alternative improved control circuit, which eliminates the above problems, is shown in Fig. 6-22. Resistor R_E degenerates the transconductance of the transistor, making the transfer function independent. This circuit also has the feature of acting as a "preload,"

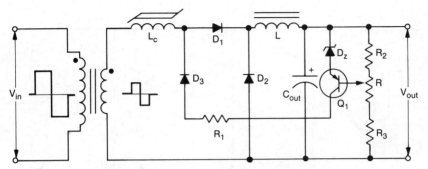

FIGURE 6-21   A transistor-based magnetic-amplifier control circuit.

that is, it prevents the magnetizing current of the reactor from raising the output voltage at no-load conditions.

The feedback network Z_f and Z_i can be designed, using techniques discussed in Chap. 9, to stabilize the loop (see Example 6-8).

Full-wave saturable core regulator circuits are also simple to design. Figure 6-23 shows how a circuit can be used to control a full-wave output, such as the outputs found in push-pull or bridge-type converters. Notice that only one control circuit is used to reset the two separate saturable cores.

6-5-4   The UC1838 Magnetic-Amplifier Controller

Unitrode Corporation has introduced the UC1838 single-chip integrated circuit, which provides all the necessary parts to implement a high-performance magnetic amplifier with or without independent current-limit control. Figure 6-24 shows the block diagram of the UC1838.

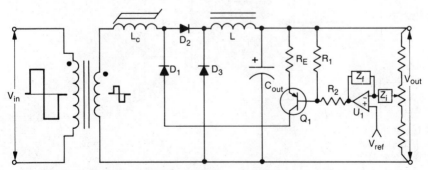

FIGURE 6-22   An improved magnetic-amplifier control circuit.

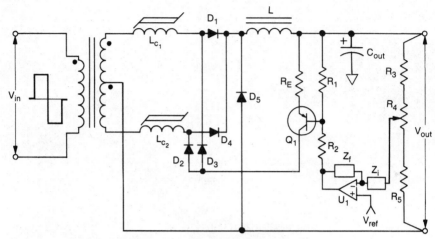

FIGURE 6-23 A full-wave magnetic-amplifier control circuit.

Hence, the integrated circuit includes the following basic functions:

1. An independent, precise, 2.5-V reference. This reference is a band gap design, accurate to within 1 percent, operating from a supply voltage of 4.5 to 40 V.

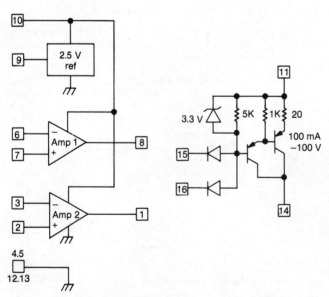

FIGURE 6-24 The UC1838 mag-amp control IC. (*Courtesy Unitrode Corporation.*)

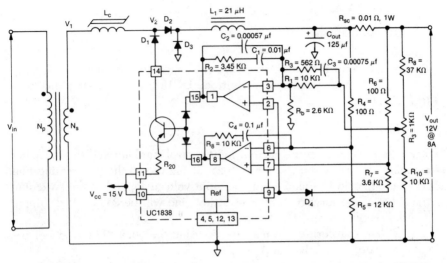

FIGURE 6-25 A magnetic-amplifier control circuit, with current limiting using the UC1838 control, for Example 6-8.

2. Two identical, high-gain operational amplifiers. The amplifiers have an input range of -0.3 V to V_{CC} and a current-sink capability of greater than 1 ma, with a slew rate of 0.3 V-μs. The amplifiers have a gain bandwidth of 800 kHz and may be cascaded if greater loop gain is required.

3. A high-voltage PNP reset current driver. The driver has the capability of delivering up to 200 ma of reset current with an 80-V collector voltage swing. With internal emitter degeneration, the reset drive operates as a transconductance amplifier providing a reset current as a function of input voltage. A mag-amp circuit using the UC1838 IC is shown in Fig. 6-25. The circuit also provides independent current-limit control.

The following example gives the reader enough information to design the feedback components of the magnetic-amplifier controller. In order to understand the concepts presented here, a knowledge of Chap. 9 material is necessary, since extensive reference is made to the K-factor techniques for the amplifier loop analysis.

EXAMPLE 6-8

Consider the forward converter shown in Fig. 6-19. The following specifications apply:

AC input: 90 to 135 V or 180 to 265 V

Outputs: 5 V @ 20 A, 12 V @ 8 A ($I_{min} = 1$ A)

Switching frequency: 100 kHz

Design the control loop op amp, for the magnetic amplifier of Example 6-7, at the 12-V output, using a Unitrode UC1838 magnetic-amplifier control IC.

SOLUTION

The schematic of both the current-limit control amplifier and the mag-amp amplifier control is shown in Fig. 6-25. Using the off-the-line rectification scheme depicted in Fig. 2-1, the dc input voltage applied to the switching transistor of the forward converter at low line will be 252-52 V dc ripple and the rectifier drops = 200 V dc.

Calculate transformer primary turns, using the 3622-PL00-3C8 Ferroxcube pot core, which has an $A_c = 2.02$ cm².

$$N_p = \frac{(V_{in,min})10^8}{2fB_{max}A_c} = \frac{200 \times 10^8}{2 \times 100 \times 10^3 \times 1.8 \times 10^3 \times 2.02} = 28 \text{ turns}$$

In order to get a secondary voltage $V_1 = 40$ V, as shown in Fig. 6-25, we need the 12 V, secondary turns, N_{s2}, to be

$$n = \frac{N_p}{N_{s2}} = \frac{V_{in,min}}{40}; \qquad N_{s2} = 40 \frac{N_p}{V_{in,min}} = 6 \text{ turns}$$

The filter inductor L_1 calculation is based on the maximum off time. The value of L_1 is calculated as follows. From Eq. 6-13:

$$t_{off} = \frac{1 - (V_{out}/V_1)}{2f} = \frac{1 - (12/40)}{2 \times 100 \times 10^3} = 3.5 \ \mu s$$

From Eq. 6-14:

$$L_1 = \frac{V_{out}(t_{off})}{(0.25)I_{out}} = \frac{12(3.5 \times 10^{-6})}{(0.25)8} = 21 \ \mu H$$

If we choose an output ripple voltage of 0.2 V, the required output capacitance may be calculated using Eq. 6-21:

$$C_{out} = \frac{I_{out}}{8f(\Delta V_{out})} = \frac{2}{8 \times 100 \times 10^3 \times 0.2} = 12.5 \ \mu F$$

The ESR of this capacitor must be

$$\text{ESR} = \frac{0.2}{12} = 0.016 \ \Omega$$

This is a very low ESR; therefore, we have to adjust the value of the capacitor to achieve the desired results. We will use 10 times the calculated capacitance, that is, 125 μF, by paralleling tantalum capacitors to achieve the desired value. Consider that a Unitrode UC1524A PWM integrated circuit is used. In this IC, a control voltage V_c is compared with a sawtooth ramp voltage V_s (2.5 V) to establish the PWM drive to transistor Q_1. For the forward converter, only one of the two alternating outputs of the IC is used, in order to limit the duty cycle δ to 50 percent maximum and therefore allow for transformer core reset. Hence,

$$\delta = \frac{0.5V_c}{V_s} = \frac{0.5V_c}{2.5} = \frac{V_c}{5}$$

Since the forward converter is a member of the buck regulator family, the output voltage is related to both the input voltage and the duty cycle by the equation

$$V_{out} = \frac{(V_{in,max})\delta}{n} = \frac{(V_{in,max})V_c}{n2V_s}$$

To derive an expression for the gain of the 12-V output, the above formula is differentiated with respect to V_c:

$$Gain = \frac{V_{in,max}}{n2V_s} = \frac{380}{(4.67)5} = 16.3$$

or

$$Gain_{dB} = 20 \log 16.3 = 24.2 \text{ dB}$$

The corner frequency of the 12-V output filter is

$$f_c \doteq \frac{1}{2\pi\sqrt{L_1 C_{out}}} = \frac{1}{6.28\sqrt{21 \times 25 \times 10^{-12}}} = 3.1 \text{ kHz}$$

Figure 6-26 shows the Bode plots of the output filter transfer function and its phase shift.

Choosing a type 3 amplifier (see Chap. 9) and the unity gain crossover frequency to be one-fifth the switching regulator's clock frequency (that is, 20 kHz), we can determine the desired gain and phase boost of the feedback amplifier. From Fig. 6-26 by inspection, at 20 kHz the gain of the modulator is 0.7 or -3 dB and the phase shift is 158°.

As recommended in Chap. 9, at least 60° of phase margin is required at the crossover frequency. Therefore the phase boost from Eq. 9-43 is

$$Boost = M - P - 90 = 60 - (-158) - 90 = 128°$$

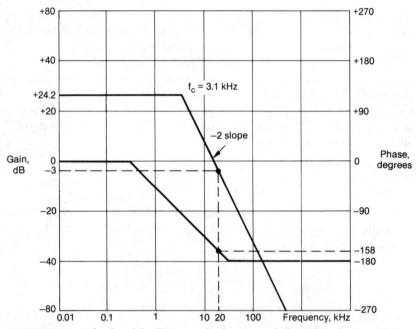

FIGURE 6-26 Bode plot of the filter transfer function and phase shift, for Example 6-8.

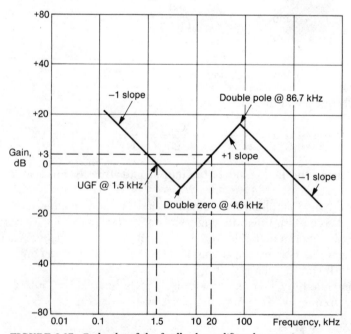

FIGURE 6-27 Bode plot of the feedback-amplifier characteristics.

The gain at the unity gain crossover frequency of 20 kHz is the reciprocal of the modulator's gain, that is, $+3$ dB or $G = 1.4$ (see Fig. 6-27).

Using Eqs. 9-49 through 9-54 and assuming $R_1 = 10$ kΩ, the required feedback-amplifier components are calculated as follows:

$$K = \left\{ \tan\left[\left(\frac{\text{boost}}{4}\right) + 45 \right] \right\}^2 = \left\{ \tan\left[\left(\frac{128}{4}\right) + 45 \right] \right\}^2 = 18.8$$

$$C_2 = \frac{1}{2\pi f G R_1} = \frac{1}{(6.28)20 \times 10^3 \times 1.4 \times 10 \times 10^3} = 0.00057 \ \mu\text{F}$$

$$C_1 = C_2(K - 1) = 0.00057(18.8 - 1) = 0.01 \ \mu\text{F}$$

$$R_2 = \frac{\sqrt{K}}{2\pi f C_1} = \frac{\sqrt{18.8}}{(6.28)20 \times 10^3 (0.01)10^{-5}} = 3.45 \ \text{k}\Omega$$

$$R_3 = \frac{R_1}{(K - 1)} = \frac{10 \times 10^3}{(18.8 - 1)} = 562 \ \Omega$$

$$C_3 = \frac{1}{2\pi f \sqrt{K} R_3} = \frac{1}{(6.28)20 \times 10^3 \sqrt{18.8}(562)} = 0.00075 \ \mu\text{F}$$

$$\text{UGF} = \frac{1}{2\pi R_1(C_1 + C_2)} = \frac{1}{(6.28)10 \times 10^3 (0.0106) \times 10^{-6}} = 1.5 \ \text{kHz}$$

Figure 6-27 shows the plot of the amplifier transfer function. The double-zero and double-pole frequency location is

$$f_z = \frac{f}{\sqrt{K}} = \frac{20 \times 10^3}{\sqrt{18.8}} = 4.6 \ \text{kHz}$$

and

$$f_p = f\sqrt{K} = (20 \times 10^3)\sqrt{18.8} = 86.7 \ \text{kHz}$$

Figure 6-27 shows the Bode plot of the feedback amplifier. The gain bandwidth product of this amplifier is

$$\text{GBW} = KGf = (18.8)(1.4)(20,000) = 526.4 \ \text{kHz}$$

Since the GBW of the UC1838 is specified at 800 kHz, the requirements of the design can be met with this amplifier. The overall system loop gain Bode plot is shown in Fig. 6-28.

The current-limit section is implemented by using a series-sensing resistor of 0.01 Ω, 1 W, precision wirewound. This circuit will provide adequate overload protection. In circuits with high current outputs, the additional efficiency loss in the sensing resistor may be avoided by providing a current-sensing transformer to drive the input of the current-limit amplifier. In this

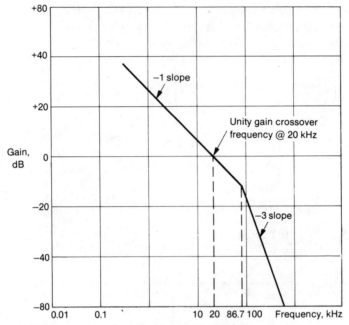

FIGURE 6-28 Overall system loop gain Bode plot.

case the current-sensing transformer should be placed between point V_2 and the anode of the diode D_2.

6-6 DESIGNING THE OUTPUT FILTER CAPACITOR

The choice of the output filter capacitor depends upon the type of converter being used as well as maximum operating current and switching frequency. Most of today's applications use electrolytic capacitors, preferably of the low ESR type. The ESR of the filter capacitor has a direct effect on the output ripple and also on the life of the capacitor itself. Since the ESR is a dissipative element, the power loss in it generates heat, which in turn shortens the capacitor's life.

Modern-day capacitors have temperature ratings of 105°C and very low ESR at frequencies above 20 kHz. As converter operating frequencies start to increase, the majority of capacitor manufacturers offer low ESR electrolytic capacitors with guaranteed performance at 100 kHz. With advancement in passive component technology, the trend at higher than 50-kHz frequencies is the development of film type capacitors which will offer a high current capability. Film type capacitors exhibit extremely low ESR, and they behave much better than electrolytics. Already some capacitor manufacturers claim film capacitor current capabilities of 2 A/μF at a working frequency of 100 kHz and above.

Regardless of the type of capacitor used for output filtering, the following analysis pertains to the calculation of its value. Referring to Fig. 6-15c, the current waveform in the output capacitor C_{out} is centered about zero and has an amplitude of ΔI. Notice that the current waveform crosses the zero reference in the positive direction at t_1, which is the middle of the on time, while at t_2, which is the middle of the off time, it crosses the zero reference in the negative direction. Thus the current will produce a ripple voltage ΔV which is given by Eq. 6-20.

$$V_{\text{out}} = \frac{1}{C_{\text{out}}} \int_{t_1}^{t_2} i \, dt \qquad (6\text{-}20)$$

But the average current during the time interval t_1 and t_2 is $(\Delta I_{\text{out}}/2)/2$ or $\Delta I_{\text{out}}/4$. Therefore, integrating Eq. 6-20 yields

$$V_{\text{out}} = \frac{I_{\text{out}}}{4C_{\text{out}}} \frac{T}{2} = \frac{(\Delta I_{\text{out}})T}{8C_{\text{out}}} = \frac{\Delta I_{\text{out}}}{8fC_{\text{out}}}$$

where T is the total period of the on time t_1 and the off time t_2.

Rearranging terms, the minimum output capacitance is

$$C_{\text{out}} = \frac{\Delta I_{\text{out}}}{8f\Delta V_{\text{out}}} \qquad (6\text{-}21)$$

where $I_{\text{out}} = 0.25 \, I_L$; $I_L = $ specified load current
$\Delta V_{\text{out}} = $ allowable peak-to-peak output voltage ripple
$f = $ operating frequency

In order to ensure minimum output voltage ripple, the ESR of the capacitor may be calculated by the following relationship:

$$\text{ESR}_{\text{max}} = \frac{\Delta V_{\text{out}}}{\Delta I_{\text{out}}} \qquad (6\text{-}22)$$

It is important to note at this point that proper selection of the LC filter is essential, since it influences two important parameters in the performance of a switching power supply. First, the LC filter combination has a very strong influence on the overall stability of the switching system, as we will see in Chap. 9. Second, a small L and large C will result in a low surge impedance of the output filter, which means that the power supply will have a good transient response due to load step changes.

In fact, measuring the transient response of a switching power supply, the important factor may not be how long it takes the output to recover during a step load change, but how deeply it deviates from the output voltage nominal value. For example, a 5-V output may not be suitable for TTL if it dips more than 250 mV during, say, a 25 percent load change, provided that this load change is anticipated in the actual application.

EXAMPLE 6-9

Calculate the capacitance value and the ESR value for an output filter capacitor to be used in the converter specified in Example 6-5, allowing a maximum output ripple of 100 mV.

SOLUTION

Using Eq. 6-21 we get

$$C_{out} = \frac{5}{8 \times 20 \times 10^3 \times 0.1} = 0.3125 \times 10^3 = 312.5 \ \mu F$$

From Eq. 6-22

$$ESR_{max} = \frac{0.1}{5} = 0.02 \ \Omega$$

Although the output capacitance of 312-μF minimum may be adequate in theory, practical experience shows that it takes a higher value to accomplish the required specifications. In fact, a rule of thumb of 300 μF/A minimum at 20 kHz is a more realistic value when electrolytic capacitors are used.

Using more than two capacitors in parallel to achieve the required capacitance and reducing the ESR to extremely low values is also recommended. In any case, careful final circuit measurements and refining of the design during prototyping will always yield optimum results, and the formulas presented above will give the designer a first-order approximation, which is good for a start.

REFERENCES

1. Archer, W. R.: Current-Driven Synchronous Rectifier, *Motorola TMOS Power FET Design Ideas*, 1986.

2. Hiramatsu, R. et al.: Switch Mode Converter Using High Frequency Magnetic Amplifier, *Powerconversion International*, March-April 1980.

3. Hnatek, E. R.: "Design of Solid State Power Supplies," 2d ed., Van Nostrand Reinhold, New York, 1981.

4. Jamerson, C.: "Calculation of Magnetic Amplifier Post Regulator Voltage Control Loop Parameters," HFPC Proceedings, April 1987.

5. Magnetics, Inc.: Output Regulators Using Mag Amp Control for Switched-Mode Power Supplies, *Powerconversion International*, 1985.

6. ———: "Molypermalloy Core Catalog."

7. Middlebrook, R. D., and S. Ćuk: "Advances in Switched Mode Power Conversion," vols. 1 and 2, Teslaco, Pasadena, Calif., 1981.

8. Mullett, C. E.: "Performance of Amorphous Materials in High-Frequency Saturable Reactor Output Regulators," HFPC Proceedings, May 1986.

9. ———, and R. Hiramatsu: "Recent Advances in High Frequency Mag Amps," HFPC Proceedings, April 1987.

10. Patel, R.: "Using Bipolar Synchronous Rectifiers Improves Power Efficiency," The Power Sources Conference, 1984.

11. ———: "Circuit Description of a Synchronous PWM Regulator," Unitrode Corp. Application Note, 1984.

12. Pressman, A. I.: "Switching and Linear Power Supply, Power Converter Design," Hayden, Rochelle Park, N.J., 1977.

13. Unitrode Corp.: "Linear Integrated Circuits Databook," 1987.

14. ———: Application Note U-68 A.

15. Watson, J. K.: "Applications of Magnetism," Wiley, New York, 1980.

SEVEN
SWITCHING REGULATOR
CONTROL CIRCUITS

7-0 INTRODUCTION

The majority of today's switching power supplies are of the pulse-width-modulated (PWM) type. This technique varies the conduction time of the switching transistor during the on period to control and regulate the output voltage to a predetermined value. Although other methods may be used for control and regulation, the PWM method offers excellent performance, such as tight line and load regulation, and stability during temperature variations.

In recent years a number of integrated circuits have been developed, which include all the necessary functions to design a complete switching power supply with the addition of few external components. The purpose of this chapter is to introduce the reader to some of the techniques and circuits used to implement the PWM control section of the switching power supply and to explain how this control is achieved.

7-1 ISOLATION TECHNIQUES OF SWITCHING REGULATOR SYSTEMS

The role of an off-the-line regulated switching power supply is twofold. First it must derive well-regulated low-level output voltages, capable of powering electronic or electromechanical circuits and devices, and second it must have a high input-to-output isolation in order to protect the user from shock hazard due to high voltage or leakage currents.

Figure 7-1 depicts two different block diagrams, showing how line isolation may be achieved in an off-the-line switching power supply. Blocks sharing common ground are depicted with identical ground symbols. These block diagrams are universal, and they may be used for any basic type of switching power supply design, such as half-bridge, full-bridge, flyback, forward, etc.

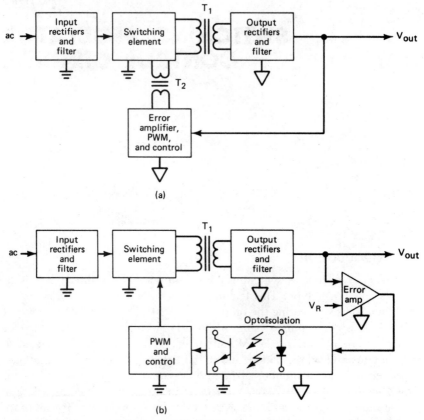

FIGURE 7-1 (*a*) Transformer isolation and (*b*) optoisolation techniques used in off-the-line switching power supplies.

In the block diagram of Fig. 7-1*a*, the error-amplified, PWM, and control circuit have a common ground with the output rectifiers and filter. Input-to-output isolation is achieved at the power transformer T_1 and at the driver transformer T_2. In general, transformer T_2 is a base or gate driver. In the block diagram of Fig. 7-1*b* the control circuit and PWM have a common ground with the switching element and the input rectifiers and filter. Input-to-output isolation is achieved at the power transformer T_1 and through the optoisolator.

Both line isolation techniques shown in Fig. 7-1 give very good perform-ance characteristics through careful circuit design. Choosing one circuit vs. the other is primarily based on economics and the type of switching power converter design. In general, the transformer isolation circuit shown in Fig. 7-1*a* can be used with all different power converter designs, while the optoisolation version shown in Fig. 7-1b is most commonly used in flyback and forward type converter designs.

7-2 PWM SYSTEMS

Although many switching techniques can be employed to implement a switched-mode power supply, the fixed-frequency PWM technique is by far the most popular choice. In a PWM system a square wave pulse is normally generated to drive the switching transistor on or off. By varying the width of the pulse, the conduction time of the transistor is accordingly increased or decreased, thus regulating the output voltage.

The PWM control circuit may be single-ended, capable of driving a single transistor converter, such as a flyback or forward. If two or more transistors have to be driven, as with half-bridge or full-bridge circuits, a dual-channel PWM circuit is necessary.

7-2-1 A Single-Ended, Discrete Component, PWM Control Circuit

A very simple closed loop, PWM control circuit may be implemented by using a small number of discrete components and semiconductor circuits, as depicted in Fig. 7-2. The function of the circuit is as follows. A clock pulse generator IC_1 generates an asymmetric square wave output at a fixed frequency, i.e., 20 kHz. This generator may easily be designed using a 555 timer, or equivalent circuitry.

The square wave is differentiated by capacitor C_1 and resistor R_1 to produce a sawtooth waveform which is used to turn off the normally conductive transistor Q_1. The resulting negative going pulse at the collector of transistor Q_1 is inverted by transistor Q_2, producing a positive going pulse at the collector of Q_2.

The low-impedance output driver combination of Q_3 and Q_4 is used to switch the main switching transistor Q_5 on or off, thus transferring energy to the output of the converter through the transformer-choke T_1. Regulation is achieved by comparing a portion of the output voltage, derived by voltage divider R_9 and R_{10}, against a fixed reference voltage V_{ref}. Any changes at the output due to line or load variations are amplified by op-amp IC_3 which drives the photodiode of optocoupler IC_2, modulating its light intensity, thus forcing phototransistor IC_2 to conduct harder. Consequently the square wave pulse at the base of transistor Q_1 is differentiated even more strongly, causing transistors Q_1, Q_2, and Q_4 to be on for a longer period of time, while transistors Q_4 and Q_5 are switched on for a shorter period. Thus the pulse width is modulated according to load and line conditions, stabilizing the output voltage.

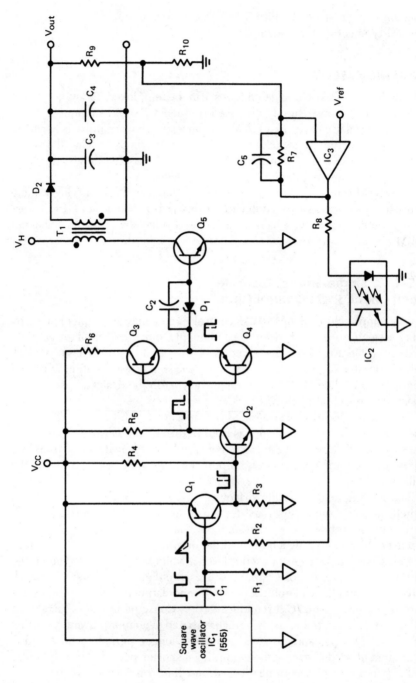

FIGURE 7-2 A PWM control circuit for a flyback switching power supply.

The circuit shown in Fig. 7-2 is greatly simplified, and refinements may be necessary when used in an actual off-the-line switching power supply application.

7-2-2 An Integrated PWM Controller

In recent years, a number of integrated circuits have been developed which include all the functions necessary to build a PWM switching power supply in a single package with just the addition of a few external components. Figure 7-3 depicts the basic building blocks of a simple PWM controller and its associated waveforms. The circuit functions as follows. An op-amp compares the feedback signal from the output of the power supply to a fixed reference voltage V_{ref}. The error signal is amplified and fed into the inverting input of a comparator. The noninverting input of the comparator accepts a sawtooth waveform with a linear slope, generated by a fixed-frequency oscillator. The oscillator output is also used to toggle a flip-flop, producing square wave outputs Q and $\overline{Q}$.

The comparator square wave output and the flip-flop outputs are both used to drive the AND gates, enabling each output when both inputs to the gate are "high." The result is a variable duty cycle pulse train at channels A and B. Figure 7-3b shows how the output pulse width is modulated when the error signal changes its amplitude, as depicted by the dotted lines. Normally the outputs of the PWM controller are externally buffered to drive the main power switching transistors. This type of circuit may be used to drive either two transistors or a single transistor. In the latter case the outputs may be externally ORed, or only one channel may be used as a driver.

The merits of such a PWM controller are profound, including the programmable fixed-frequency oscillator, linear PWM section with duty cycle form 0 to 100 percent, adjustable dead time to prevent output transistor simultaneous conduction, and above all simplicity, reliability, and cost-effectiveness.

7-3 SOME COMMERCIALLY AVAILABLE MONOLITHIC PWM CONTROL CIRCUITS AND THEIR APPLICATIONS

In the early 1970s the switching power supply market started to expand in the commercial sector, a fact which brought about the first major attempt by integrated circuit manufacturers to offer PWM control circuits in a single chip. The first such circuits to appear in the market were the Motorola MC3420 Switchmode Regulator Control Circuit and the Silicon General SC3524 PWM Control Circuit, which was destined to become the industry standard.

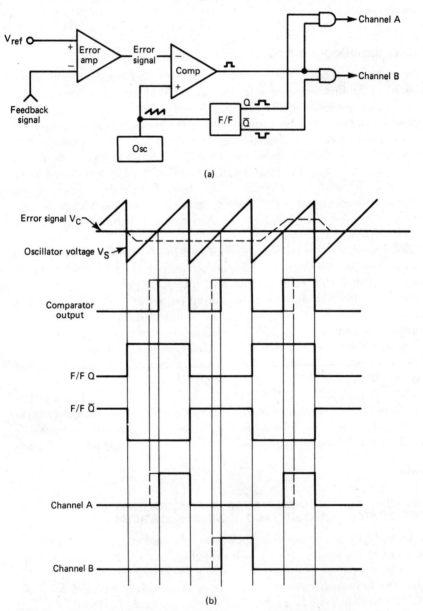

FIGURE 7-3 (a) An integrated PWM control circuit and (b) its associated waveforms.

These PWM controllers were and are the heart for a complete switching power supply design and may be used equally well in single-ended or dual-channel applications. Shortly thereafter manufacturers started to introduce more PWM control circuits with improved characteristics and more features. The Texas Instruments TL494 PWM Control Circuit is an improved version of the SG3524, offering features such as adjustable dead-time control, output transistors with high source or sink capabilities, improved current limiting control, output steering control, etc.

With the introduction of the power MOSFET, the first PWM control circuits appeared with totem-pole outputs, capable of driving MOSFETs directly as well as bipolars, such as the SG1525A and SG1526 series. Besides offering all the features found in previous control circuits, these new ICs included additional features such as undervoltage lockout, programmable soft start, digital current limiting, and operation up to 400 kHz.

Although, as mentioned, all the above circuits may be used in all the popular switch-mode topologies, recently some companies have introduced PWM controllers optimized for high efficiency in forward or flyback power converters. One such circuit is the Motorola MC34060 PWM controller, which includes all the features needed to implement a forward or flyback design using a minimum of external parts.

Another circuit is the Unitrode UC1840 series, which promises to have a major impact on the design of single-ended power converters. The PWM controller includes all the control, driving, monitoring, and protection functions needed to build a complete switching power supply with just the addition of a few passive external components. Features of the controller include a low-current, off-line start circuit; built-in protection from over-voltage, undervoltage, and overcurrent conditions; feed-forward line regulation over a 4:1 input range; 500-kHz operation, etc.

In the next paragraphs we describe the function of some of the available PWM control integrated circuits in order to familiarize the reader with the operation of these circuits. The descriptions presented are purely informational, and the reader is advised to review carefully all available data sheets from different manufacturers of PWM control circuits in order to choose the optimum IC controller for his or her particular application.

7-3-1 The TL494 PWM Control Circuit

The TL494 is a fixed-frequency PWM control circuit, incorporating all the building blocks necessary for the control of a single-ended or dual-channel switching power supply. Figure 7-4 shows the internal construction and block diagram of the TL494 controller. An internal linear sawtooth oscillator is frequency-programmable by two external components, R_T and C_T, con-

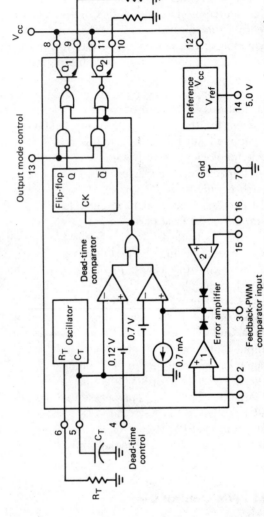

FIGURE 7-4 Internal block diagram of the TL494 PWM controller. (*Courtesy of Motorola Semiconductor Products, Inc.*)

nected on pins 6 and 5, respectively. The oscillator frequency is determined by

$$f_{osc} = \frac{1.1}{R_T C_T} \tag{7-1}$$

Output pulse-width modulation is accomplished by comparison of the positive sawtooth waveform across capacitor C_T to either of two control signals. The NOR gates, which drive output transistors Q_1 and Q_2, are enabled only when the flip-flop clock-input line is in its low state. This happens only during that portion of time when the sawtooth voltage is greater than the control signals. Therefore, an increase in control-signal amplitude causes a corresponding linear decrease of output pulse width, as depicted by the timing diagram waveforms of Fig. 7-5.

The control signals are external inputs that can be fed into the dead-time control pin 4, the error amplifier inputs at pins 1, 2, 15, and 16, or the 120-mV input offset which limits the minimum output dead time to approximately the first 4 percent of the sawtooth cycle time. This would result in a maximum duty cycle of 96 percent with the output mode control pin 13 grounded, and 48 percent with the same pin connected to the reference line. Additional dead time may be imposed on the output by setting the dead-time control input pin 4 to a fixed voltage, ranging from 0 to 3.3 V.

The PWM comparator provides a means for the error amplifier to adjust the output pulse width from the maximum percent on time, established by the dead-time control input, down to zero, as the voltage at the feedback

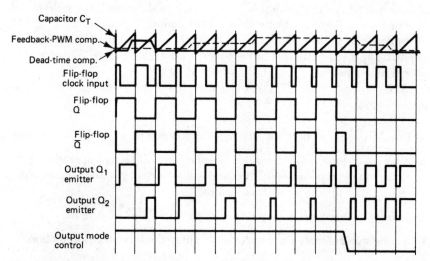

FIGURE 7-5 Timing diagram waveforms of the TL494 PWM controller. (*Courtesy of Motorola Semiconductor Products, Inc.*)

pin varies from 0.5 to 3.5 V. Both error amplifiers have a common-mode input range from -0.3 to $(V_{CC} - 2)$ V and may be used to sense power supply output voltage and current.

The error amplifier outputs are active high and are ORed together at the noninverting input of the PWM comparator. With this configuration, the amplifier that demands minimum output on time dominates control of the loop.

When capacitor C_T is discharged, a positive pulse is generated on the output of the dead-time comparator, which clocks the pulse steering flip-flop and inhibits the output transistors Q_1 and Q_2. With the output mode control pin 13 connected to the reference line, the pulse steering flip-flop directs the modulated pulses to each of the two output transistors alternately for push-pull operation. The output frequency is equal to half that of the oscillator.

Output drive can also be taken from Q_1 or Q_2, when a single-ended operation with a maximum duty cycle of less than 50 percent is required. When higher output currents are required for single-ended operation, Q_1 and Q_2 may be connected in parallel, and the output mode control pin must be tied to ground to disable the flip-flop. The output frequency will now be equal to that of the oscillator.

Figure 7-6 demonstrates the use of the TL494 in a simple PWM push-pull converter with current limiting protection.

7-3-2 The UC1840 Programmable, Off-Line PWM Controller

Although most of the commercially available PWM controllers are designed for universal usage, the Unitrode UC1840 family of programmable PWM controllers was specifically designed for application as a primary-side, cost-effective approach to flyback or feed-forward designs. Figure 7-7 shows the overall block diagram of the UC1840 PWM controller.

In reference to Fig. 7-7, the UC1840 contains the following distinct features:

1. Fixed-frequency operation, user-programmable by a simple RC network

2. A variable slope ramp generator for constant volt-second operation, providing open loop line regulation and minimizing, or in some cases eliminating, the need for feedback control

3. A drive switch for low-current start-up, with direct off-line bias

4. A precision reference generator with internal overvoltage protection

5. Complete undervoltage and overcurrent protection including programmable shutdown and restart

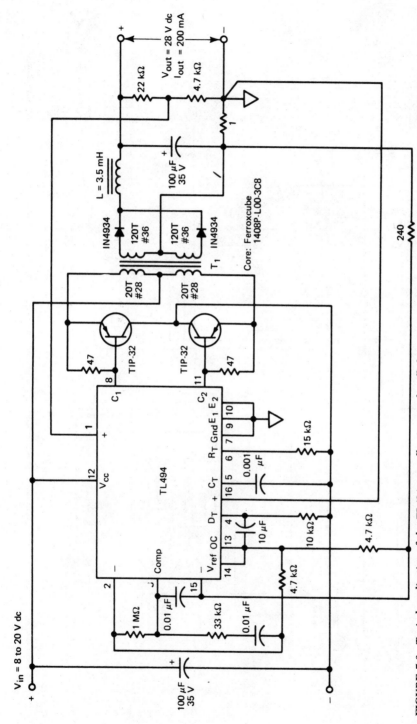

FIGURE 7-6 Typical application of the TL494 controller in a push-pull, low-current power converter with short-circuit protection. (*Courtesy of Motorola Semiconductor Products, Inc.*)

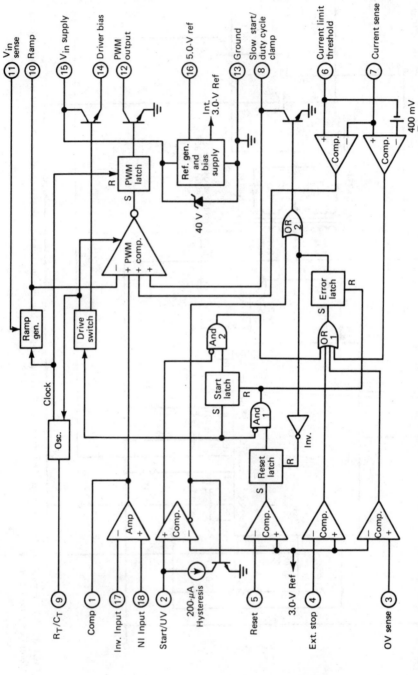

FIGURE 7-7 Block diagram of the UC1840 PWM integrated controller, optimized for primary side control of off-the-line switching power supplies. (*Courtesy of Unitrode Corporation.*)

6. A high-current, single-ended PWM output optimized for fast turn-off of an external power switch

7. Logic control for pulse-commandable or dc power sequencing

The following is a discussion of how the UC1840 PWM control circuit functions. References to Figs. 7-7 and 7-8 are made throughout the presentation. During the initial power-up, and before the voltage at pin 2 reaches 3 V, the start/undervoltage (UV) comparator pulls a current of 200 μA, causing an added drop across resistor R_4. At the same time the drive switch holds the driver bias transistor off, ensuring that the only current required through resistor R_{in} is the start-up current. Also the slow start transistor is on, holding pin 8 of the IC low, thus keeping capacitor C_S discharged.

The start latch flip-flop keeps the undervoltage signal UV from being defined as a fault. The start voltage level is defined as

$$V_C \text{ (start)} = 3 \left(\frac{R_4 + R_5}{R_5} \right) + 0.2R_4 \qquad (7\text{-}2)$$

When the control voltage rises above that level, the start/UV comparator eliminates the 200-μA hysteresis current, sets the start latch flip-flop to monitor for an undervoltage fault, activates the driver bias output transistor to supply base current to the power switch, and also turns off the slow start transistor, providing soft-start start-up which is set by R_S and C_S.

Pin 8 of the UC1840 may be used for both soft-start turn-on and duty cycle limiting, as well as a PWM shutdown port. The duty cycle may vary from 0 to 90 percent, and maximum duty cycle limiting is achieved by clamping the voltage on pin 8 with a divider formed by resistors R_S and R_{DC}. When fixed ramp slope operation is employed, resistor R_S is taken to the 5-V reference. Alternatively, for a constant volt-second operation, i.e., the ramp generator connected as shown in Fig. 7-5, R_S must be connected to the dc input line.

The desired maximum duty cycle is set by the voltage at pin 8, defined by

$$V \text{ (pin 8)} = \left(\frac{R_{DC}}{R_S + R_{DC}} \right) V_{DC,in} \qquad (7\text{-}3)$$

This clamping voltage must be equal to the ramp voltage, at the same dc input voltage level.

The ramp generator on the other hand will produce an output ramp voltage with a slope given by

$$\frac{dV}{dt} = \frac{V_{line}}{R_R C_R} \qquad (7\text{-}4)$$

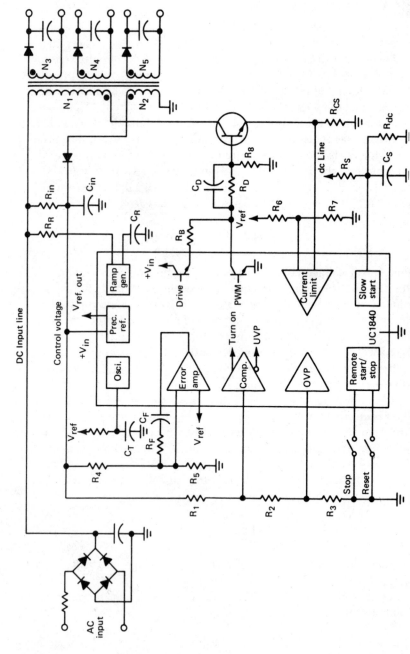

FIGURE 7-8 A typical off-the-line isolated flyback power supply, centered around the UC1840 PWM controller. The power supply is fully protected against undervoltage (UV), overvoltage (OV), and short circuit, and features soft start and self-bias. (*Courtesy of Unitrode Corporation.*)

where V_{line} is the voltage to where resistor R_R is connected. For a fixed ramp slope, R_R should be connected to the 5-V reference. The peak voltage of the ramp is clamped at 4.2 V, while its valley or low voltage is typically 0.7 V.

The PWM section of the UC1840 consists of the oscillator, the ramp generator, the error amplifier, the PWM comparator, the PWM latch flip-flop, and the PWM output transistor, as depicted in the block diagram of Fig. 7-7. The function of the PWM section is conventional. A constant clock frequency is established by connecting a simple RC network from pin 9 to ground and the 5-V reference as shown in Fig. 7-8. The frequency of oscillation is

$$f = \frac{1}{R_T C_T} \qquad (7\text{-}5)$$

where R_T can range from 1 to 100 kΩ and C_T from 300 pF to 0.1 μF.

The basic function of the ramp generator was described above. The error amplifier is a voltage-mode op-amp with a common-mode range of 1 to $(V_{in} - 2)$ V. Thus, either input of the op-amp may be directly connected to the 5-V reference. The other input of the amplifier is sensing an equivalent portion of the output (or input) voltage which is to be controlled.

The ramp generator output, the error amplifier output, as well as the slow start input and the current limit output, are inputted to the PWM comparator. The comparator generates the output pulse which starts at the termination of the clock pulse and ends when the ramp waveform crosses the lowest of the three positive inputs to the comparator. The clock develops a blanking pulse to keep the duty cycle below 100 percent. The PWM latch flip-flop ensures one pulse per period rate and eliminates oscillation at comparator crossover. The PWM output pulse arrives at pin 12 of the UC1840, an open collector transistor. The output transistor is capable of supplying 200 mA of output current; thus, it can drive directly bipolars or MOSFETs. If higher output current is required, external buffering may easily be implemented. Ancillary circuits such as overvoltage sense, external stop, and reset are easily implemented.

Current limiting and overcurrent shutdown are implemented with comparators of different thresholds. In the event of an overload, these comparators shorten the PWM output pulse and at the same time turn on the slow start transistor, discharging the soft-start capacitor, ensuring proper restart at the end of the fault.

7-3-3 The UC1524A PWM Control Integrated Circuit

The UC1524A PWM control IC is an advanced version of the first commercially developed PWM controller, the SG1524. Since many examples contained in this book are referenced to the UC1524A, the following description of its building blocks will adequately introduce the reader to its function. For more in-depth specifications, the individual data sheets should be consulted. The block diagram of the UC1524A integrated PWM control IC is shown in Fig. 7-9.

An internal linear sawtooth oscillator is frequency programmable by a resistor R_T and a capacitor C_T. The oscillator frequency is determined by

$$f_{osc} = \frac{1.15}{R_T C_T} \tag{7-6}$$

and it is usable to frequencies beyond 500 kHz.

The ramp voltage swings approximately 2.5 V to change the comparator output from 0 to 1, by comparing it to either one of two control signals, i.e., the error amplifier output or the current-limit amplifier output. The error amplifier input range extends beyond 5 V, eliminating the need for a pair of dividers, for 5-V outputs. Note also that the UC1524A has an on-board 5-V reference which is trimmed to a ±1 percent accuracy.

Output pulse-width-modulation is accomplished by steering the resulting modulated pulse out of the high-gain comparator to the PWM latch along with the pulse steering flip-flop which is synchronously toggled by the oscillator output.

The PWM latch insures freedom from multiple pulsing within a period, even in noisy environments. In addition, the shutdown circuit feeds directly to this latch, which will disable the outputs within 200 ns of activation. The current-limit amplifier is a wide-band, high-gain amplifier, which is useful for either linear or pulse-by-pulse current limiting in the ground or power

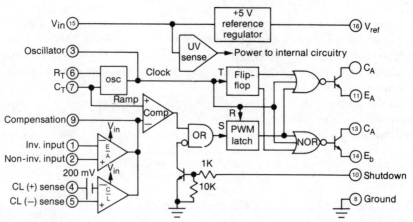

FIGURE 7-9   Block diagram of the UC1524A PWM control IC. (*Courtesy Unitrode Corporation.*)

supply output lines. Its threshold is set at 200 mV. An undervoltage lockout circuit has been added, which disables all the internal circuitry, except the reference, until the input voltage is 8 V. This action holds standby current low until turn-on, greatly simplifying the design of low-power, off-line switchers.

The power capability of the output transistors is 200 mA, and their voltage rating is 60 V. The transistors may be paralleled for increased current capacity.

As it can be seen, this versatile IC controller can be used in a variety of isolated or nonisolated switching power supplies for a number of applications. A simple buck regulator application is shown in Fig. 7-10.

In this application, a wide input range nonisolated buck regulator is presented, with the addition of a small signal 2N2222 transistor which serves to provide a constant drive current to the output switch, regardless of the input voltage level. Note the simplicity of the current-limit function, which can be implemented using a small sensing resistor in series with the output bus.

Figure 7-11 shows a low-cost, 50-W off-the-line, fully isolated forward converter. At power on, initial start-up is provided by C_2 which charges

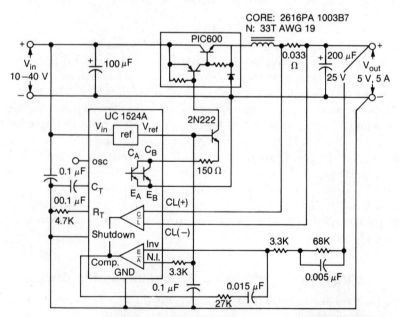

FIGURE 7-10 The UC1524A and PIC600 hybrid switch provides a wide input range, 25-W buck regulator, with a minimum number of external components. (*Courtesy Unitrode Corporation.*)

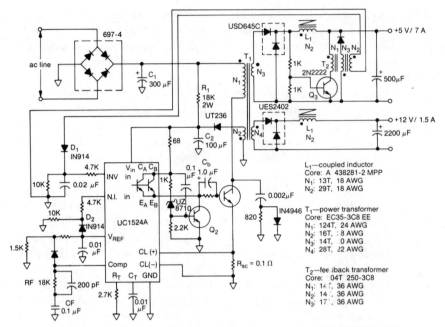

FIGURE 7-11 An off-the-line isolated forward converter using the UC1524A. (*Courtesy Unitrode Corporation.*)

through R_1. Winding N_2 takes over after steady-state conditions have been reached, providing the drive power. Isolated feedback control is provided by an innovative circuit comprised of transistor Q_3 and transformer T_2, by sampling the 5-V output level at the switching frequency of 40 kHz.

With every switching cycle, the output voltage is transferred from winding N_1 to winding N_2 where it is peak detected to generate a primary reference signal to drive the PWM error amplifier. Diode D_2 is used to temperature compensate for the loss in rectifier D_1, and the net result is better than 1 percent regulation.

Diode D_3 forms a duty cycle clamp. Forward base drive current to the switching transistor Q_1 is provided by connecting the internal drive transistors in parallel. Turn-off base current is provided by the combination of transistor Q_2 and capacitor C_b.

7-3-4 The UC1846 Current-Mode Control Integrated Circuit

In Chap. 3 the basic current-mode topology was discussed, and the advantages over conventional PWM techniques were presented.

The Unitrode Corporation UC1846 is an integrated current-mode control circuit, which has all the necessary functions to design state-of-the-art, high-frequency switching power supplies, with the addition of a handful of external components.

Figure 7-12 shows the block diagram of the UC1846. The integrated circuit provides the following nine features:

1. A ±1 percent, 5.1-V trimmed band gap reference used both as an external voltage reference and internal regulated power source to drive low-level circuitry.

2. A fixed frequency sawtooth oscillator with variable dead-time control and external synchronization capability. Circuitry features on all NPN design capable of producing low-distortion waveforms in excess of 1 MHz. Figure 7-13 shows the implementation of the oscillator circuit.

   An external resistor R_T is used to generate a constant current into a capacitor C_T to produce a linear sawtooth waveform. The oscillator frequency is given by

$$f_{\text{osc}} = \frac{2.2}{R_T C_T} \tag{7-7}$$

   where $1\ \text{k}\Omega \le R_T \le 500\ \text{k}\Omega$ and $C_T \ge 1000\ \text{pF}$.

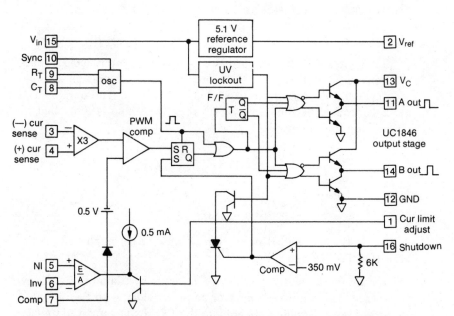

FIGURE 7-12 **The block diagram of the UC1846 current-mode control integrated circuit.** (*Courtesy Unitrode Corporation.*)

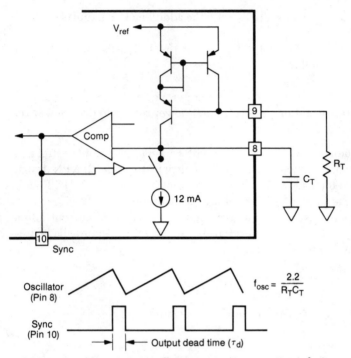

FIGURE 7-13 **The UC1846 oscillator circuit.** (*Courtesy Unitrode Corporation.*)

Referring to Fig. 7-13, the oscillator generates an internal clock pulse used, among other things, to blank both outputs and prevent simultaneous cross conduction during switching transitions. The output dead time is controlled by the oscillator fall time. Fall time is controlled by C_T according to the formula

$$t_d = 145C_T \left[\frac{12}{12 - (3.6/R_T)} \right] \qquad (7\text{-}8)$$

For large values of R_T

$$t_d = 145C_T \qquad (7\text{-}9)$$

Synchronization of more than one circuit may be accomplished via the bidirectional SYNC pin.

3. An error amplifier with common mode range from ground to 2 V below bias voltage. The output of the error amplifier along with the output of the current sense amplifier are fed into the PWM comparator to provide the pulse steering modulated signal.

4. A current sense amplifier which may be used in a variety of ways to sense peak switch current for comparison with the error voltage. Referring to Fig. 7-12, maximum swing on the inverting input of the PWM comparator is limited to approximately 3.5 V by the internal regulated supply. Accordingly, for a fixed gain of 3, maximum differential voltages must be kept below 1.2 V at the current sense inputs. Figure 7-14 depicts several ways of configuring sense schemes.

Direct resistive sensing is simplest; however, a lower-peak voltage may be required to minimize power loss in the sense resistor. An RC filter is recommended to eliminate any extraneous voltage spikes which may trip the current sense latch and result in erratic operation of the PWM circuit, particularly at low duty cycles.

Transformer coupling can provide isolation and increase efficiency at the cost of added complexity. Regardless of the scheme, the largest sense voltage consistent with lower power losses should be chosen for noise immunity. Typically, this will range from several hundred millivolts in some resistive sense circuits to the maximum of 1.2 V in transformer coupled circuits.

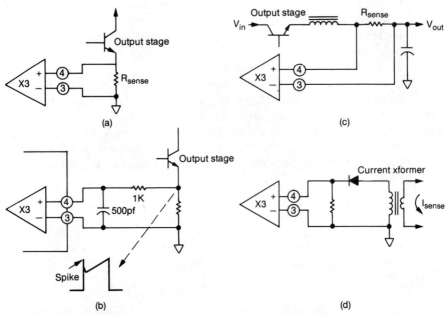

FIGURE 7-14 Various current sense schemes for the UC1846 current sense amplifier: (*a*) Resistive sensing with ground reference, (*b*) same sensing with RC filter to reduce switch transient spikes, (*c*) resistive sensing above ground, and (*d*) isolated current sensing. (*Courtesy Unitrode Corporation.*)

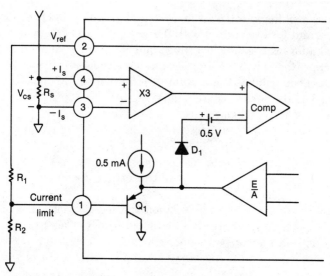

FIGURE 7-15 Peak current-limit design for the UC1846. (*Courtesy Unitrode Corporation.*)

5. Current limiting through clamping of the error signal at a user-programmed level. In fact, one of the most attractive features of a current-mode converter is its ability to limit peak switch currents on a pulse-by-pulse basis by simply limiting the error voltage to a maximum value. Figure 7-15 shows how this is accomplished.

Resistors R_1 and R_2 form a divider network to set a predetermined voltage at pin 1. This voltage, in conjunction with transistor Q_1, acts to clamp the output of the error amplifier at a maximum value. Since the base-emitter drop of Q_1 and the forward drop of diode D_1 nearly cancel, the negative input of the comparator will be clamped at the voltage at pin 1 minus 0.5 V. Thus the differential input voltage, V_{cs}, of the current sense amplifier is

$$V_{cs} = \frac{V_{\text{pin1}} - 0.5}{3} \tag{7-10}$$

Using this relationship, a value for maximum switch current in terms of external programming resistors can be derived as follows:

$$I_{cl} = \frac{[R_2(V_{\text{ref}})/(R_1 + R_2)] - 0.5}{3R_s}$$

Note also that R_1 supplies holding current for the shutdown circuit and therefore should be selected prior to selecting R_2.

6. A shutdown function with built-in 350-mV threshold. May be used in either a latching or nonlatching mode, or initiating a "hiccup" mode of operation. The shutdown circuit is shown in Fig. 7-16. Shutdown is accomplished by applying a signal greater than 350 mV to pin 16. In this configuration, capacitor C_s is used to provide a soft-start or soft-restart function, if so desired.

7. Under voltage lockout with hysteresis to guarantee that the outputs will stay "off" until the reference is in regulation.

8. Double pulse suppression logic to eliminate the possibility of consecutively pulsing either output.

9. Totem pole output stages capable of sinking or sourcing 100 mA continuous, 400-mA peak currents.

All the above features indicate that the UC1846 current-mode control IC is a versatile circuit which can be used in many designs.

Figure 7-17 shows how the UC1846 could be utilized in a conventional push-pull circuit. This is a 20-kHz power supply, and although no effort is given to matching the switching transistors, there is no current imbalance to cause transformer saturation. For higher operating frequencies, the switching transistors may be replaced by MOSFETs. The power supply will also exhibit excellent transient response to output load changes, as compared to conventional PWM designs.

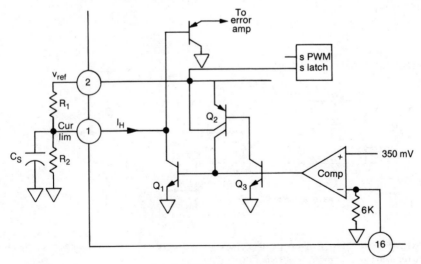

FIGURE 7-16 Shutdown circuit of the UC1846. (*Courtesy Unitrode Corporation.*)

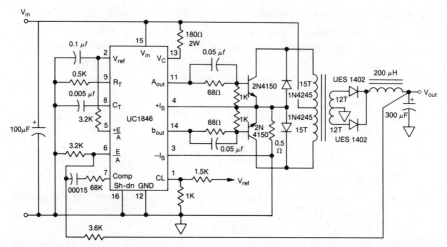

FIGURE 7-17   A 20-kHz push-pull current-mode controlled converter using the UC1846. (*Courtesy Unitrode Corporation.*)

Also by disabling the oscillator and error amplifiers (C_T grounded, $+E/A$ to V_{ref} and $-E/A$ grounded) of one or more slave modules, and connecting SYNC and COMP pins of the slave(s), respectively, similar outputs may be paralleled together to provide a modular approach to power supply design. In this configuration, output current will be equally shared by all units. Figure 7-18 shows a practical implementation of slaving two units in a parallel operation.

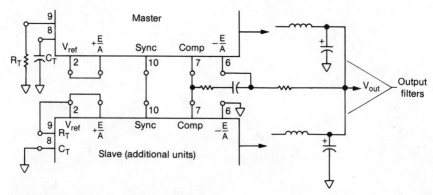

FIGURE 7-18   Slaving allows parallel operation of two or more UC1846 units, with equal current sharing. (*Courtesy Unitrode Corporation.*)

7-3-5 The UC1860 Resonant-Mode Power Supply Controller

Increased interest in resonant-mode power supplies has been fueled by the fact that this class of converters is well-suited for operation in the megahertz region. Although a number of discrete power components have been developed to operate in the megahertz region, the lack of similar simple integrated controllers has been obvious. Unitrode Corporation, however, is in the process of releasing the UC1860 resonant-mode power supply controller family, which will satisfy this need. Information given here is preliminary, since at the writing of this book, only advanced product information was available, but it should serve very well as an introduction to this versatile IC. The block diagram of the UC1860 controller is depicted in Fig. 7-19.

The control philosophy employed is fixed on-time, variable frequency. The fundamental control blocks include a reference and a wide band, high-gain error amplifier which controls a variable frequency oscillator from 1

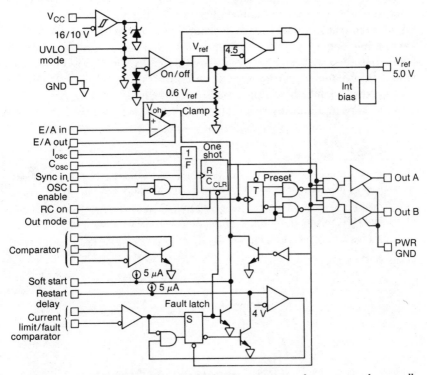

FIGURE 7-19 The block diagram of the UC1860 resonant mode power supply controller. (*Courtesy Unitrode Corporation.*)

kHz to 3 MHz. The error amplifier controls oscillator frequency via a resistor into the I_{osc} pin which is 2 diode drops above ground. The error amplifier is clamped so that its output swing is limited from 2 diode drops above ground to 2 V plus 2 diode drops, thus allowing minimum and maximum frequencies to be programmed.

A temperature-stable one-shot timer, triggered by the oscillator, generates pulses as low as 300 ns defining on-time for the output drivers. Each output is capable of driving transient currents up to 2 A, making them ideal for power MOSFET gates. The mode of the toggle flip-flop is programmable for alternate or unison operation of the outputs.

Additional blocks reside in the chip to facilitate more control capability. An uncommitted open-collector comparator can be used to shorten the on-time pulse during start-up or under low load conditions. A fast comparator with a common-mode range from -0.3 to $+3$ V is available to sense over current fault conditions. The chip is versatile in fault disposition with several soft-start and restart delay options. The restart delay pin can be used to permanently shut the supply down after a fault, restart after a delay, or restart immediately after the fault indication has been removed.

A programmable undervoltage lockout (UVLO) rounds out the IC. It allows off-line operation with a 16-V start threshold and 6 V to V_{cc} hysteresis or operation directly from a dc supply from 5 to 20 V. The UVLO mode pin can also be used as a part for gating the entire supply on and off. During undervoltage lockout, the output stages are actively driven low, and supply current is kept to a minimum.

REFERENCES

1. Mamano, B.: "Applying the UC1840 to Provide Total Control for Low Cost, Primary-Referenced Switching Power Systems," Unitrode, Application Note U-91, 1982.

2. ———— and R. Patel: "The UC1524A Integrated PWM Control Circuit Provides New Performance Levels for Old Standard," Unitrode, Application Note U-90, 1982.

3. Motorola, Inc.: "Linear/Switchmode Voltage Regulator Handbook," 2d ed., 1982.

4. Unitrode Corp.: "Linear Integrated Circuits Databook," 1987.

EIGHT

SWITCHING POWER SUPPLY ANCILLARY, SUPERVISORY, AND PERIPHERAL CIRCUITS AND COMPONENTS

8-0 INTRODUCTION

In general, switching power supplies are closed-loop systems, a necessity for good regulation, ripple reduction, and system stability. Other than the basic building blocks, which have been described in the previous chapters, there are a number of peripheral and ancillary circuits that enhance the performance and reliability of the power supply.

Components such as optoisolators are extensively used by primary referenced designs like the flyback or feed-forward converter to offer the necessary input-to-output isolation and still maintain good signal transfer information. Other circuits, such as soft-start, overcurrent, and overvoltage protection circuits, are used to guard the power supply against failures due to external stresses. This chapter describes the components used in these circuits, and presents typical circuit designs showing how these devices may be utilized to perform their function.

8-1 THE OPTICAL COUPLER (OR OPTOISOLATOR)

The optical coupler, or optocoupler, is used primarily to provide isolation between the input and output of the power supply, while at the same time output noise and low temperature coefficient of 50 ppm/°C are desirable of a typical optocoupler.

The optocoupler consists mainly of two elements: the light source, which could be an incandescent lamp or a light-emitting diode (LED); and the detector, which could be a photovoltaic cell, photodiode, phototransistor, or light-sensitive SCR. The most common construction of the optocoupler consists of a gallium arsenide (GaAs) LED and a silicon phototransistor

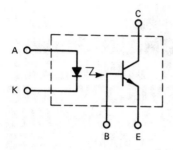

FIGURE 8-1 A typical optocoupler circuit.

housed in a light-excluding package. In normal operation, a current through the LED produces a light source, the intensity of which depends on the exciting current, which in turn modulates the phototransistor producing a collector current proportional to the forward diode current. Figure 8-2 shows an optocoupler connected in a basic linear operational mode.

In order to design the input circuit, the parameters needed are the diode forward current I_F, the diode forward voltage V_F, and the input voltage V_{in}. Then the current limit resistor R may be calculated from the following equation:

$$R = \frac{V_{in} - V_F}{I_F} \tag{8-1}$$

In general, diode forward voltage is plotted vs. diode forward current and is given in the manufacturer's specification sheets. Thus, an operating

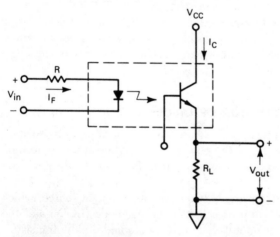

FIGURE 8-2 An optocoupler connected in a basic linear mode. A diode forward current I_F generates a light source, inducing collector current I_C to flow in the phototransistor.

point may be chosen and the current limit resistor can be easily calculated. The basic parameter of interest in the output section of the coupler is the collector current I_C of the phototransistor.

The amount of current generated in the collector of the phototransistor is proportional to the diode forward current I_F and the optocoupler's dc current transfer ratio or coupling efficiency η. If the diode forward current is known, the collector current of the phototransistor may be calculated by

$$I_C = \eta I_F \tag{8-2}$$

The manufacturer's data sheets normally give a dc current transfer ratio curve at a specified collector-emitter voltage V_{CE}. From this information the collector current (and emitter current) may be derived, which allows choosing the value of R_L in order to get a specified output voltage V_{out} (see Fig. 8-2).

8-2 A SELF-BIAS TECHNIQUE USED IN PRIMARY SIDE-REFERENCED POWER SUPPLIES

Although base drive transformers may be used to provide the necessary input-to-output isolation in a switching power supply, they are most commonly used in bridge type designs, mainly because of necessity rather than choice. The majority of the flyback or forward converter designs use an optoisolator to achieve the necessary isolation.

The use of an optoisolator greatly simplifies a design, because it eliminates the need for the driver transformer and the bias transformer. Since, in this case, the total control circuitry may be primary side-referenced, start-up and self-bias techniques may be used directly from the high-voltage line and the high-frequency transformer to bias the control circuitry.

Figure 8-3 shows the implementation of a self-bias circuit which may be used in this family of power supplies. The function of the circuit is as follows. When the ac input voltage is applied, the PWM control and drive circuit receive a bias voltage V_C set by the linear regulator formed by R_1, Z_1, and Q_1 connected directly to the dc high-voltage bus. After the power supply starts, an auxiliary winding in the main transformer provides a voltage V_D, the value of which is chosen by design higher than the voltage V_C, consequently back-biasing diode D_5 and turning off the linear regulator. Under this condition the power supply provides a self-bias voltage V_D to keep it running, while no power is dissipated in the start-up regulator.

Caution must be exercised when designing this circuit to select a high-voltage transistor capable of sustaining the base-collector high-voltage stress when it is off. This circuit is a typical application of a self-exciting auxiliary power supply, and many other circuits based on this principle may be developed to suit individual needs.

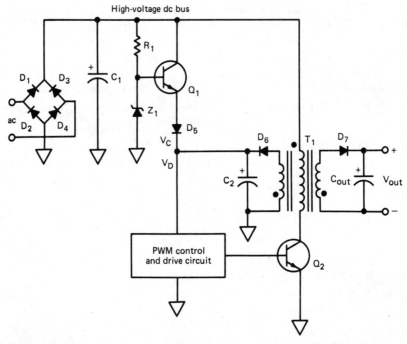

FIGURE 8-3 Start-up and self-bias circuitry used in primary side-referenced, off-the-line switching power supplies.

8-3 OPTOCOUPLER CIRCUIT DESIGN TO PROVIDE INPUT-TO-OUTPUT ISOLATION IN A SWITCHING POWER SUPPLY

When the optocoupler is used in an off-the-line switching power supply for the purpose of providing input-to-output isolation, the following design criteria must be kept in mind:

1. The optocoupler must sustain an isolation breakdown voltage as dictated by local and/or international safety standards.

2. The amplifier circuitry driving the optocoupler must be well designed to compensate for the coupler's thermal instability and drift.

3. An optocoupler with good coupling efficiency is preferred.

Normally, the optocoupler in such an application is used in a linear mode; that is, a control voltage at the input of the coupler results in a proportional output voltage to be used for further control, such as closed-loop regulation.

A typical circuit connected in this mode of operation is depicted in Fig. 8-4. The function of this circuit is as follows. A portion of the output voltage

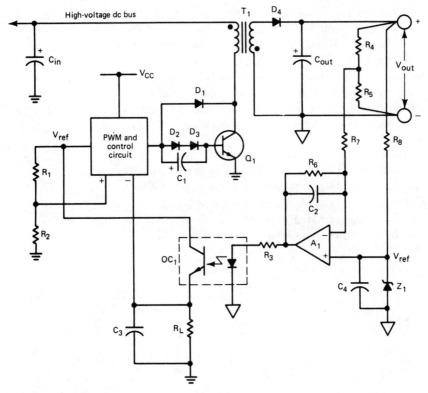

FIGURE 8-4 Regulation control, as well as input-to-output isolation, is provided in this flyback circuit using an optocoupler.

in this flyback circuit is derived by the voltage divider comprised of resistors R_4 and R_5. This voltage is fed into one input of the amplifier A_1 and is compared against a fixed reference voltage V_{ref}, which is fed into the other input of amplifier A_1.

Any difference between the two voltages is amplified, and in turn the voltage at the output of the amplifier A_1 produces a current through R_3, consequently modulating the light intensity of the coupler's LED. The LED light source induces a proportional emitter current at the phototransistor, and therefore a voltage drop is developed across R_L, which mirrors the voltage at the junction of R_4 and R_5.

The voltage across R_L is fed into the noninverting input of the error amplifier of the PWM circuit (see Fig. 7-3), while the inverting input of the error amplifier is sitting at a fixed voltage, derived by dividing down a reference voltage source V_{ref}. Therefore the conduction period of transistor Q_1 is adjusted accordingly, in order to maintain regulation at the output of the power supply.

Although the circuit shown in Fig. 8-1 is a practical realization, the number of components needed to drive the LED of the optocoupler may be dramatically reduced using the circuit shown in Fig. 8-5. The function of this circuit is exactly the same as the one described for Fig. 8-3. Notice here the simplicity and the reduction of parts by using the TL431 shunt regulator. This shunt regulator is offered by a number of companies, i.e., Texas Instruments, Motorola, etc. It can be used as a programmable, low-temperature-coefficient, reference amplifier with current sink capabilities up to 100 mA.

The internal 2.5-V reference makes it ideal for operation from a 5-V bus, while its output voltage may be externally programmed up to 36 V. Its low output noise and low temperature coefficient of 50 ppm/°C are desirable features for such an application. Figure 8-6 shows the symbol and block diagram of the TL431. In Fig. 8-5, the C_2-R_6 capacitor and resistor network may be necessary in an actual application for frequency compensation.

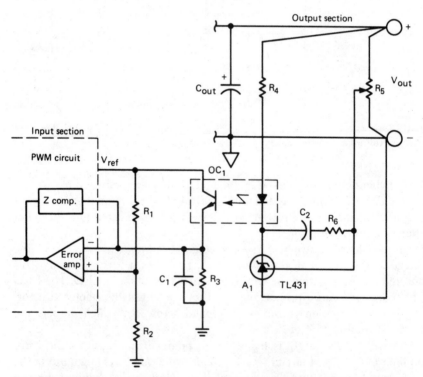

FIGURE 8-5 Using the TL431 shunt regulator to drive the optocoupler LED and to provide the necessary amplification function dramatically reduces component count.

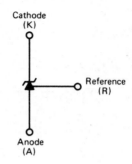

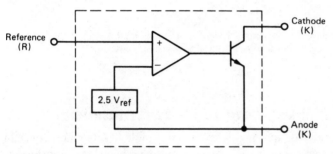

FIGURE 8-6 Symbol and block diagram of the TL431 programmable precision reference. (*Courtesy of Motorola Semiconductor Products, Inc.*)

Another inexpensive, low parts count circuit with good performance may be designed around a single transistor as shown in Fig. 8-7. In this circuit transistor Q_1 is biased at a set level using zener diode Z_1. The collector current flow excites the photodiode and sets the control voltage level across R_L.

Output voltage adjust is accomplished by varying potentiometer R_{out}, thus modulating the light intensity of the photodiode. The R_1-C_1 low-frequency filter is added to improve overall power supply stability.

All the above-described optocoupler driving circuits are typical examples, and modifications may be necessary to tailor them to fit individual needs. On the other hand these circuits may be used to develop other circuitry, even to be used in specialized applications. The fact remains that the optocoupler is an important peripheral device in the design of switching power supplies, providing input-to-output isolation and still maintaining all the regulation features of the converter.

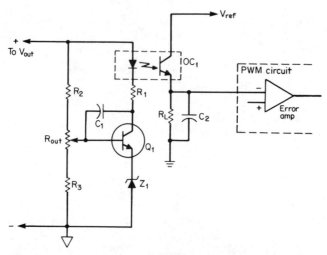

FIGURE 8-7 A single-transistor amplifier-comparator may be used to drive the optocoupler in the feedback loop of a switching power supply.

8-4 SOFT START IN SWITCHING POWER SUPPLY DESIGNS

In most switching power supply designs it is desirable to introduce a certain delay during start-up, in order to avoid output overshoots and transformer saturation problems at turn-on. Circuits which are employed to perform this task are called *soft-start circuits*, and in general they consist of an RC network which allows the PWM control circuit output to increase from zero to its operating value very "softly."

Figure 8-8 shows how a soft-start circuit may be implemented in a PWM control circuit. At time $t = 0$, when the power supply is just turned on, capacitor C is discharged and the error amplifier output is held to ground through diode D_1, thus inhibiting the comparator output.

At time $t = 0^+$, the capacitor starts to charge through resistor R with a time constant determined by

$$\tau = RC \tag{8-3}$$

toward the charging voltage V_{ref}. As capacitor C attains full charge, diode D_1 is back-biased, and therefore the output of the error amplifier is isolated from the soft-start network. The slow charge of capacitor C results in the gradual increase of the PWM waveform at the output of the comparator, and consequently a "soft start" of the switching element is initiated.

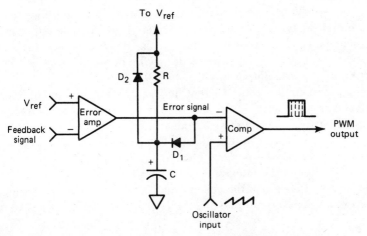

FIGURE 8-8 A typical soft-start circuit used in a PWM control circuit aids
the gradual increase of the PWM signal to its operating value.

Diode D_2 is used to bypass resistor R in order to discharge the capacitor
C fast enough in case of system shut-down, thus initiating a new soft-start
cycle even during very short interrupt periods. In some second-generation
PWM control integrated circuits, R has been substituted by an on-chip
current generator, thus the only external element required to implement
the soft-start feature is the addition of capacitor C.

It is obvious that the soft start imposes a certain delay on the rise time
of the output voltage, thus a reasonable value of R and C must be selected
to keep this delay within practical limits.

8-5 CURRENT LIMIT CIRCUITS

A switching power supply is generally designed to operate safely at a predetermined output power level. Operation beyond the nominal output current should be avoided, but in case an overcurrent or short-circuit condition
occurs, the power supply must have some means of protection against per-
manent destruction.

Current limit circuits are basically protective networks which will limit
the output current drain to a safe level in case of an output short-circuit
condition. There are many ways of implementing a current limit circuit by
placing it either at the primary (input) side of the power supply or at the
output section. Of course, the optimum current limit topology greatly depends on the specific power supply design which it protects. Single-output

designs may be equally protected by placing the current limit at either the input or the output section. Thus, for primary-referenced direct drive power supplies, it may be convenient to place the current limit at the input side, while for power designs using base drives, it may be beneficial to place the current limit circuit at the output bus.

Although direct coupling of the current limit circuit with the monitored bus may be convenient, simple, and may require a low parts count, transformer-coupled current limit circuits are also extensively used, especially when grounds are not common and voltage level translation is required. Current limit circuits may be implemented by discrete components, or in IC PWM control circuits the integral current limit function may be used.

It should be noted here that the current limit must have a fast response in order to protect the power supply before destruction occurs.

8-5-1 Current Limit Circuits for Primary-Referenced Direct Drive Designs

Primary-referenced direct drive designs, such as flyback or forward converters, are easily current limited. Figure 8-9 shows two ways of achieving current limiting in such designs.

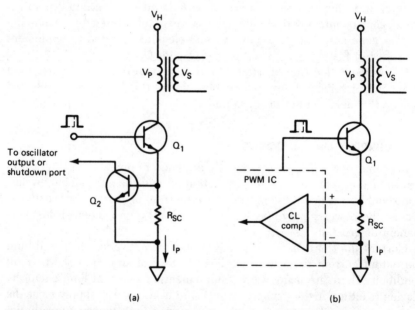

FIGURE 8-9 In flyback or forward converters, a simple resistor in series with the switch will provide the necessary voltage drop to turn on the transistor (*a*), or activate the IC comparator (*b*), and shorten the drive pulse in case of overcurrent conditions.

In Figure 8-9a, the peak primary current is monitored in terms of a proportional voltage drop developed across the current limit resistor R_{SC}. The value of resistor R_{SC} is given by

$$R_{SC} = \frac{V_{BE}}{I_P} \qquad (8\text{-}4)$$

When the voltage drop across resistor R_{SC} exceeds the base-emitter threshold voltage, transistor Q_2 turns on. The collector of Q_2 is returned to the oscillator output or a shut-down port.

If an overload or short exists at the output of the power supply, primary current increases dramatically, turning transistor Q_2 on, harder and harder. The collector of Q_2, in turn, pulls the oscillator output toward ground, or it activates the shutdown circuit, thus limiting the available primary current to safe levels.

A faster and more accurate current limit circuit is shown in Fig. 8-9b. Such a current limit is popular among PWM control circuit ICs. Although the principle of operation of this circuit is similar to the one described previously, there are certain distinct advantages in using this circuit vs. the transistor one. First, the comparator's current limit–activating threshold voltage is preset to an accurate and predictable level, as opposed to the wide V_{BE} threshold of a bipolar transistor. Second, this voltage threshold is made small enough, typically 100 to 200 mV, allowing the use of a smaller current limit sense resistor value, therefore increasing the overall efficiency of the converter.

8-5-2 Current Limit Circuits for Designs Utilizing Base Drivers

Normally in designs which utilize base drive isolation between the control circuitry and the switching transistor, such as half-bridge, full-bridge, and/ or flyback and forward converters, the output section shares a common ground with the control circuit. In such cases the current limit circuit may be directly connected to the output bus. Such an implementation is shown in Fig. 8-10.

Under normal operation the load current I_L is small enough to develop a sufficient voltage drop across R_{SC} to turn on transistor Q_1. Since Q_1 is off and $I_{C1} = 0$, capacitor C_1 is totally discharged and consequently transistor Q_2 is off. If I_L is increased to a value such that

$$I_L R_{SC} = V_{BE,Q1} + I_{B1} R_1 \qquad (8\text{-}5)$$

then collector current I_{C1} starts to flow, charging capacitor C_1 with a time constant determined by

$$\tau = R_2 C_1 \qquad (8\text{-}6)$$

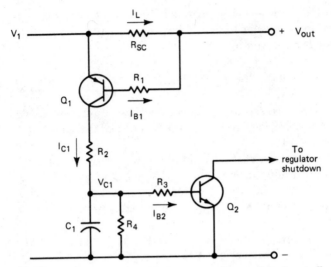

FIGURE 8-10 This current limit circuit may be used in almost all power-converted designs, where the control circuit shares the same ground with the output return bus.

The charging voltage at capacitor C_1 will attain a value

$$V_{C1} = I_{B2}R_3 + V_{BE,Q2} \tag{8-7}$$

In order to minimize loading on the capacitor voltage, a Darlington transistor with a very high h_{FE} is chosen for Q_2. This will limit the base current I_{B2} to microamperes. Choose $R_4 \ll R_3$ in order to allow fast discharge of C_1, after a current overload has been detected.

R_2 is selected as follows:

$$I_{B1,max} = \frac{V_1 - V_{BE,Q1}}{R_1}$$

and

$$I_{C1} = \beta_{Q1}I_{B1,max}$$

Therefore,

$$R_2 \geq \frac{(V_1 - V_{CE,sat,Q1})R_1}{(V_1 - V_{BE,Q1})} \tag{8-8}$$

In a properly designed circuit, V_{C1} reaches its value fast enough to bias transistor Q_2 on, which in turn shuts down the regulator's drive signal.

Recovery of the circuit is automatic upon removal of the overload. If integrated PWM control circuits are used with built-in current limit comparators, the circuit in Fig. 8-9b may very well be used by moving the current limit resistor R_{SC} at the output positive bus.

Although both methods work well to detect an overcurrent condition, the presence of the power resistor R_{SC} may become objectionable, especially in high-current outputs because of power dissipation and its adverse effect on system efficiency. If that is the case, the circuit in Fig. 8-11 may be used. This circuit uses a current transformer to detect overcurrent conditions, and since no loss elements are involved, the overall efficiency of the power supply is increased. The circuit operates as follows. A current transformer T_1 monitors load current I_L, producing a proportional voltage determined by the scaling resistor R_1. Diode D_3 rectifies the pulsed voltage, and resistor R_2 and capacitor C_1 are chosen to smooth the voltage.

When a current overload occurs, the voltage across C_1 increases to the point where Zener diode Z_1 conducts, turning on transistor Q_1. The signal at the collector of Q_1 may be used to shut down the regulator's drive signal. Current transformer T_1 may be wound on a ferrite or MPP toroid, but care must be exercised to keep the core out of saturation. Normally the primary consists of one turn, while the number of secondary turns needed to establish the necessary secondary voltage is given by

$$\frac{N_P}{N_S} = \frac{I_S}{I_P} \tag{8-9}$$

Since $I_S = V_S/R_1$, the number of secondary turns required to produce the desired voltage across the capacitor C_1 at maximum specified load current I_L is given by

$$N_S = N_P \frac{I_P R_1}{V_S + V_{D3}} \tag{8-10}$$

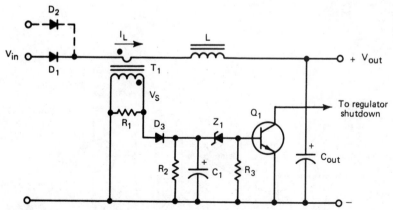

FIGURE 8-11 A nondissipative current limit circuit, using a current transformer to detect overcurrent conditions.

Eq. 8-10 gives accurate information for manufacturing the current transformer. Final adjustments in the number of turns may be made during actual circuit tests for optimum performance.

8-5-3 A Universal Current Limit Circuit

A universal current limit circuit may be designed that is equally able to perform well when placed in either the input or the output section of the power supply. Also, this circuit is well suited for multiple output voltages, where the multiplicity of outputs makes individual current limiting a difficult task.

Figure 8-12 shows the basic circuit design. This circuit shows operation at the input side of the power supply. Current transformer T_1 monitors the primary current of power transformer T_2. The secondary voltage of transformer T_1 is rectified by diodes D_1 through D_4 and smoothed by capacitor C_1. Resistor R_1 is used to set the comparator trip threshold. Under normal operation the V_{ref} terminal to the comparator is in a higher potential, and the comparator output is high. Therefore the 555 one-shot multivibrator has a low output, holding transistor Q_1 off.

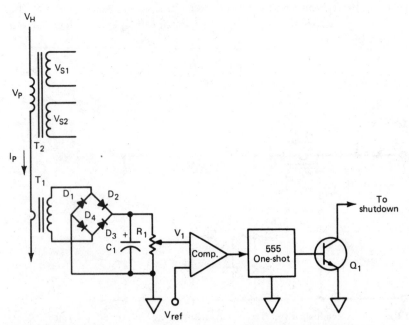

FIGURE 8-12 A one-shot multivibrator is used to initiate current limit "hiccup" when an overload is detected.

If an overload occurs, the voltage V_1 will rise above V_{ref}, pulling the output of the comparator low. The high-to-low transition at the input of the 555 produces a one-shot output, turning on transistor Q_1. The collector of the transistor is tied to the shutdown input or across the soft-start capacitor of the PWM circuit, and by pulling it toward ground it terminates the output switching pulse, shutting down the regulator.

If the overload persists, the power supply enters a hiccup mode; that is, it turns on and off, with a period determined by the 555 one-shot RC time constant, until the overload is removed. Recovery then is automatic. Design of the toroidal current transformer is the same as described in Sec. 8-5-2.

8-6 OVERVOLTAGE PROTECTION CIRCUITS

Overvoltage protection circuits are networks designed to clamp the output voltage to a safe value, should this voltage attempt to rise beyond a predetermined value. Although the threat of an overvoltage condition was quite likely in linear power supplies, this is not necessarily the case with switching power supplies. In fact, most of the failures in a switching power supply result at a "no output" condition. Why then use overvoltage protection circuits?

The reason is twofold. First, in power supplies with adjustable outputs, overvoltage protection circuits guard against accidental output overadjustment. Second, they reassure the end user about safety against overvoltage, even during the rare occasions of such a happening.

A better place to use an overvoltage protection (OVP) circuit is right at the electronic circuit which the power supply powers. An OVP circuit there will definitely protect the circuits from accidental overvoltage application caused by assembly errors, especially when more than one voltage is to be wired on the same circuit. The simplest and most effective way of implementing an OVP circuit is to use a "crowbar" SCR across the dc power bus. When an overvoltage condition is detected, the SCR is turned on by some means, shorting the output terminals. Since during turn-on the SCR is subjected to large amounts of current, careful selection of the device must be made to suit the intended application.

8-6-1 The Zener Sense OVP Circuit

One of the most widely used OVP circuits is the one shown in Fig. 8-13. Although this circuit provides very poor gate drive for the SCR and also decreases the di/dt handling capability of the SCR, it works fairly well for a low-cost design. Under normal operation, the gate of the SCR is grounded, keeping the device off. When an overvoltage is detected, zener Z_1 conducts, pulling the SCR gate to the zener voltage, thus turning it on and consequently shorting the output terminals.

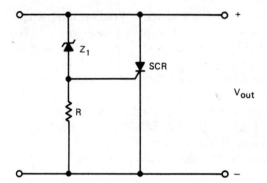

FIGURE 8-13 An OVP circuit consisting of a zener diode and an SCR.

Once fired, the SCR remains on until its anode voltage is removed. In power supplies this may be accomplished by removing the input power for a few seconds.

8-6-2 Integrated OVP Circuits

In recent years a few dedicated OVP integrated circuits have been introduced by a number of manufacturers. Most of these circuits are low-cost and offer the designer a number of design features, such as programmable trip voltage threshold, fast response, low-temperature coefficient trip point, etc.

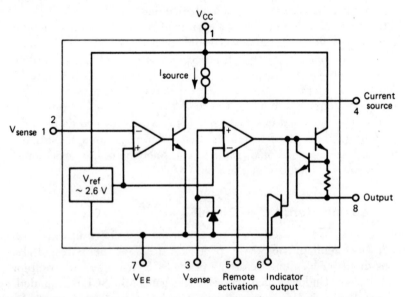

FIGURE 8-14 The MC3423 OVP circuit block diagram. (*Courtesy of Motorola Semiconductor Products, Inc.*)

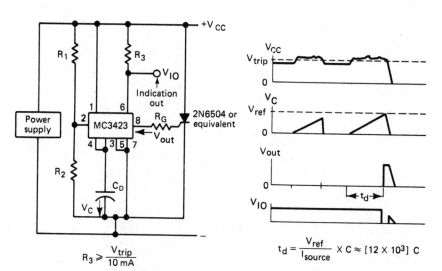

FIGURE 8-15 Typical OVP application of the MC3423. (*Courtesy of Motorola Semiconductor Products, Inc.*)

The earliest of these ICs was the MC3423, which has become an industry standard. A basic block diagram of this IC is given in Fig. 8-14. The diagram shows that the circuit consists of a stable 2.6-V reference, two comparators, and a high current output. The output is activated by a voltage greater than 2.6 V on pin 2, or by a high logic level on the remote activation, pin 5.

Figure 8-15 shows a typical application of the MC3423 in an OVP application. In the circuit, resistors R_1 and R_2 set the threshold trip voltage. The relationship between V_{trip} and R_1, R_2 is given by

$$V_{trip} = 2.6 \left(1 + \frac{R_1}{R_2} \right) \tag{8-11}$$

Keeping the value of R_2 below 10 kΩ for minimum drift is recommended.

The value of R_1 and R_2 may also be calculated by using the graph of Fig. 8-16. In the graph, $R_2 = 2.7$ kΩ, while R_1 may be directly calculated from the intersection of the trip voltage with the desired curve.

The MC3423 OVP circuit also has a programmable delay feature, which prevents false triggering when used in a noisy environment. In Fig. 8-15, capacitor C_D is connected from pins 3 and 4 to the negative rail to implement this function. The circuit operates as follows. When V_{CC} rises above the trip point set by R_1 and R_2, the internal current source begins charging the capacitor C_D connected to pins 3 and 4. If the overvoltage condition remains present long enough for the capacitor voltage V_{CD} to reach V_{ref}, the output is activated. If the overvoltage condition disappears before this occurs, the capacitor is discharged at a rate 10 times faster than the charging time,

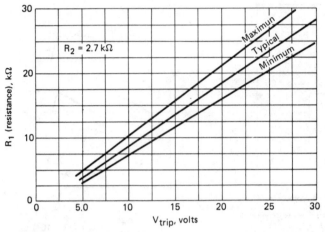

FIGURE 8-16 The threshold resistor values may be directly calculated from this graph, which plots R_1 vs. trip voltage for the MC3423 OVP circuit. (*Courtesy of Motorola Semiconductor Products, Inc.*)

resetting the timing feature. The value of the delay capacitor C_D may be found from Fig. 8-17.

A more elaborate OVP circuit, the MC3425, is in many respects similar to the MC3423, but the former one may also be programmed for undervoltage detection and also line loss monitoring. The block diagram of the MC3425 is shown in Fig. 8-18. Notice that this is a dual-channel circuit, with the overvoltage (OV) and undervoltage (UV) input comparators both referenced to an internal 2.5-V regulator. The UV input comparator has a feedback-activated, 12.5-μA current sink I_H, which is used for programming the hysteresis voltage V_H. The source resistance feeding this input R_H determines the amount of hysteresis voltage by

$$V_H = I_H R_H = (12.5 \times 10^{-6})R_H \qquad (8\text{-}12)$$

Separate delay pins 2 and 5 are provided for each channel to independently delay the drive and indicator output pins 1 and 6, respectively, thus providing greater input noise immunity. The two delay pins are essentially the outputs of the respective input comparators and provide a constant current source I_d of typically 200 μA when the noninverting input is greater than the inverting input level. A capacitor connected from these delay pins to ground will provide a predictable delay time t_d for the drive and indicator outputs. The delay pins are internally connected to the noninverting inputs of the OV and UV output comparators, which are referenced to the internal 2.5-V regulator. Therefore delay time t_d is based on the constant current source

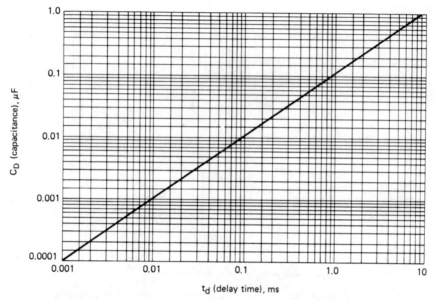

FIGURE 8-17 Delay capacitance C_D vs. minimum overvoltage duration t_D for the MC3432 OVP circuit. (*Courtesy of Motorola Semiconductor Products, Inc.*)

I_d, changing the external delay capacitor C_d to 2.5 V. The delay time t_d can be calculated as follows:

$$t_d = \frac{V_{ref}C_d}{I_d} = \frac{2.5C_d}{200} = (12.5 \times 10^3)C_d \qquad (8\text{-}13)$$

From Eq. 8-13 the delay capacitor can be easily calculated for a wide range of delay times, or it may be directly determined using the graph of Fig. 8-17.

The delay pins are pulled low when the respective input comparator's noninverting input is less than the inverting input. The sink current I_d capability of the delay pins is greater than 1.8 mA and is much greater than the typical 200 μA source current, thus enabling a relatively fast delay capacitor discharge time.

The overvoltage drive output is a current-limited emitter follower, capable of sourcing 300 mA at a turn-on slew rate of 2 A/μs, ideal for driving "crowbar" SCRs. The undervoltage indicator output is an open collector, NPN transistor, capable of sinking 30 mA to provide sufficient drive for LEDs, small relays, or shut-down circuitry. These current capabilities apply to both channels operating simultaneously, providing device power dissi-

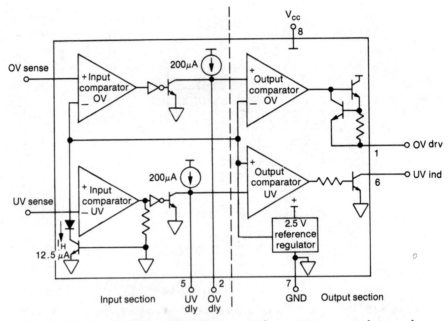

FIGURE 8-18 The dual-channel MC3425 power supply supervisory, overvoltage, under-voltage protection circuit. (*Courtesy Motorola Semiconductor Products, Inc.*)

pation limits are not exceeded. The MC3425 has an internal 2.5-V band gap reference regulator with an accuracy of ±4 percent for the basic devices.

Figure 8-19 shows a typical application of the MC3425 in an overvoltage protection circuit, with undervoltage fault indication. Note that delay capacitors have been added to pins 2 and 5.

8-7 AC LINE LOSS DETECTORS

In many computer applications, an ac line loss or a brownout condition must be detected in timely fashion in order to transfer valuable data to nonvolatile memory banks or to trigger an uninterruptible power supply (UPS) to take over before the power failure occurs. This line loss detection must be done within a cycle or two of the ac frequency, since most of the switching power supplies have a hold-up time of 16 ms minimum, enough to power the circuits between line loss detection and take-over time.

Using the MC3425 integrated circuit, the power supply designer may accomplish two design goals. First, half the MC3425 may be utilized as an OVP circuit, while the other half may be used to detect ac line loss or brownout conditions. Figure 8-20 shows a typical application of such a circuit.

As a line loss detector, input pin 4 of the MC3425 is connected as an undervoltage sensing circuit to sense the center tap of a full-wave rectified signal proportional to the ac line voltage. At each peak of the line the output of the comparator discharges the delay capacitor C_d. If a half cycle is missing from the line voltage or if a brownout occurs reducing the peak line voltage, the delay capacitor is not discharged and it continues to be charged as shown in Fig. 8-21. If a sufficient number of half cycles are missing or if the brownout continues for a sufficient time, the circuit will detect an ac line fault and will output a line fault indication on the indicator output, pulling pin 6 low.

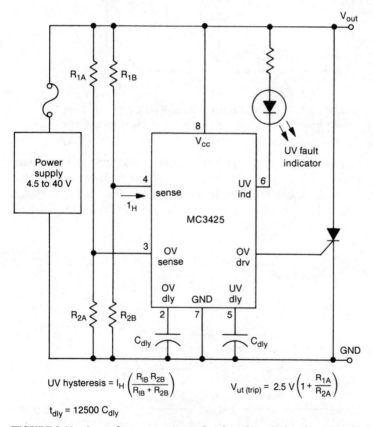

$$\text{UV hysteresis} = I_H \left(\frac{R_{1B} R_{2B}}{R_{1B} + R_{2B}} \right) \qquad V_{ut \, (trip)} = 2.5 \text{ V} \left(1 + \frac{R_{1A}}{R_{2A}} \right)$$

$$t_{dly} = 12500 \, C_{dly}$$

FIGURE 8-19 **Overvoltage protection and undervoltage fault indication with programmable delay, using the MC3425.** (*Courtesy Motorola Semiconductor Products, Inc.*)

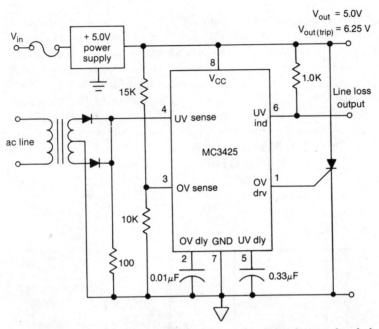

FIGURE 8-20 The MC3425 integrated circuit may be used to detect ac line faults and overvoltage conditions, independently. (*Courtesy Motorola Semiconductor Products, Inc.*)

The delay capacitor is used to provide some noise immunity and to prevent the loss of a single half cycle from triggering the line fault signal. The minimum time the fault condition must occur can be adjusted by changing the value of the delay capacitor. The graph of Fig. 8-17 may be used to specify the delay capacitor.

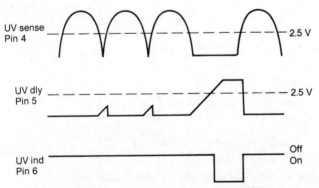

FIGURE 8-21 Waveforms illustrating brownout conditions and line loss detection for the circuit in **Fig. 8-20**. (*Courtesy Motorola Semiconductor Products, Inc.*)

REFERENCES

1. Chryssis, G. C.: Avoid Supply Voltage and Overload Problems by Defining Them Before They Happen, *Electronic Design*, August 2, 1979.

2. ———: Single Transistor Adds Latching Function to Overvoltage Protection IC, *Electronic Design*, August 16, 1979.

3. Motorola, Inc. "Linear/Switchmode Voltage Regulator Handbook," 2d ed., 1982.

4. Unitrode Corp.: "Power Supply Design Seminar Manual."

5. ———: "Linear Integrated Circuits Databook," 1987.

NINE
STABILITY IN SWITCHING POWER SUPPLIES: ANALYSIS AND DESIGN

9-0 INTRODUCTION

The subject of stability, which pertains to the closed-loop frequency response of switching regulators, has undoubtedly received much attention, and a score of papers have been published on and around the subject. To most practicing engineers as well as students, however, feedback control loop stability seems to be shrouded by a cloud of mystery. Although most designers do understand what causes a switching power supply to oscillate, many of them stabilize the loop using a trial-and-error approach or fancy mathematical models which require extensive use of the computer.

This chapter presents feedback loop stability, blending theory and practice in a coherent way, in order to give the reader the necessary tools to stabilize switching regulators with little effort, much enjoyment, and great reward.

9-1 THE LAPLACE TRANSFROM

In most linear systems, the input-to-output relationship characterizes the system, and the differential or integrodifferential equations which mathematically describe the system give an idea of the response to some input excitation. These equations are generally given in the time domain and are difficult to manipulate. By applying the Laplace transform, the equations are "transformed" to the frequency domain, thus taking an algebraic form, which is easier to handle. After the desired result has been derived, transformation back to the time domain may be achieved by applying the inverse Laplace transform.

By definition, if $f(t)$ is any function of time such that $f(t) = 0$ for $t < 0$ and the integral $\int_0^\infty f(t)e^{-st}\,dt$ has a finite value, then $f(s)$ is the Laplace transform of $f(t)$. The Laplace operator s is defined as the complex variable

$$s = \sigma + j\omega \tag{9-1}$$

and the Laplace transform is defined as

$$f(s) = \int_0^\infty f(t)e^{-st}\, dt \tag{9-2}$$

EXAMPLE 9-1

Find the Laplace transform of the unit step function, defined as $f(t) = 1$ for $t > 0$ and $f(t) = 0$ for $t < 0$.

SOLUTION

Using Eq. 9-2 we have

$$f(s) = \int_0^\infty 1e^{-st}\, dt = -\frac{1}{s} e^{-st} \Big|_0^\infty = -\frac{1}{s}(e^{-\infty} - e^0)$$

therefore

$$f(s) = \frac{1}{s}$$

It can be seen from Example 9-1 that any time function may be transformed in terms of the complex variable s. In case the result is required in the time domain, the inverse Laplace transform, given by

$$f(t) = \frac{1}{2\pi j} \int_{\sigma-j\infty}^{\sigma+j\infty} f(s)e^{st}\, ds \tag{9-3}$$

may be used to obtain $f(t)$.

There are tables which give both $f(s)$ and $f(t)$, so that transformation from time to frequency domain and vice versa may be done quickly and efficiently.

9-2 TRANSFER FUNCTIONS

So far so good, but how can the Laplace transform be used in our study of system stability to derive useful information? The first step is to derive a relationship between input driving signal and output response of a system. As an example, let us examine the simple RC network given in Fig. 9-1.

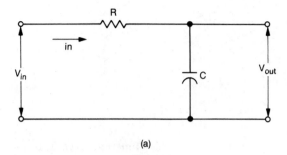

(a)

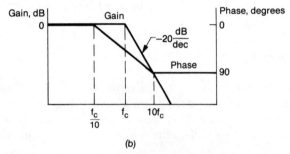

(b)

FIGURE 9-1 (*a*) An RC integrator circuit and (*b*) its gain and phase plot.

Using Kirchhoff's law the network equations may be written as

$$V_{in} = iR + \frac{1}{C} \int i \, dt$$

and

$$V_{out} = \frac{1}{C} \int i \, dt$$

Substituting $q = \int i \, dt$, the above equations reduce to

$$V_{in} = R \frac{dq}{dt} + \frac{q}{C}$$

and

$$V_{out} = \frac{q}{C}$$

Taking the Laplace transform,

$$V_{in}(s) = \left(sR + \frac{1}{C} \right) q(s) \tag{9-4}$$

and

$$V_{out}(s) = \frac{q(s)}{C} \tag{9-5}$$

Dividing Eq. 9-4 into Eq. 9-5 we get

$$\frac{V_{out}(s)}{V_{in}(s)} = \frac{1}{sRC + 1} \tag{9-6}$$

The ratio of $V_{out}(s)/V_{in}(s)$ is defined as the transfer function $G(s)$. The important thing here is to recognize that this function has a gain and a phase associated with it. Any system, therefore, may be described by its transfer function

$$G(s) = \frac{N(s)}{D(s)} \tag{9-7}$$

In such an equation, the roots of $N(s) = 0$ are called the zeros of the system, while the roots of $D(s) = 0$ are called the poles of the system. The most convenient way of plotting the gain and phase of a transfer function is on a decibel basis, and such plots are called Bode plots, after the man who developed them.

9-3 BODE PLOTS

We mentioned before that a transfer function equation has poles and zeros and that these poles and zeros determine the slope of the gain plot. Let us now examine Eq. 9-6 and Fig. 9-1. The equation shows a pole in the denominator. By setting $sRC + 1 = 0$, we get

$$sRC = -1$$

and

$$s = -\frac{1}{RC} \Rightarrow f = -\frac{1}{2\pi RC} \tag{9-8}$$

Equation 9-8 shows a very important result—that a pole will cause the transition of the gain plot from 0 to -1 at a frequency $f_c = 1/2\pi RC$. This frequency is called the *corner* or *break* frequency, so termed because the asymptote breaks here.

If we determine the rate of change of this asymptote, we see that the slope is -6 dB per octave, or -20 dB per decade. An octave is a 2:1 range of frequencies, while a decade is a 10:1 range of frequencies. Likewise the phase of the network changes at points $f_c/10$ and $10f_c$, producing a 90° phase lag.

To summarize, then, a pole will cause a transition from a $+1$ to a 0 slope, or 0 to -1, or -1 to -2, or -2 to -3, etc. This corresponds to a gain change per octave of $+6$ dB, 0 dB, -6 dB, -12 dB, and -18 dB, associated with a phase shift of $+90°$, $0°$, $-90°$, $-180°$, $-270°$, respectively. The gain change per decade is $+20$ dB, 0 dB, -20 dB, -40 dB, and -60 dB, associated with a phase shift of $+45°$, $0°$, $-45°$, $-90°$, $-135°$.

Zeros, on the other hand, are points in frequency where the slope of the Bode plot breaks upward, causing a transition of the gain plot from a -1 to 0 slope, or -2 to -1, or -3 to -2, etc. Accordingly the phase shift would now be leading by $90°$. Figure 9-2 shows circuits which cause poles and circuits which cause both poles and zeros.

In order, then, to plot any network in a Bode format, first determine its transfer function using Laplace transforms; then arrange the equation in the form

$$\frac{V_{out}}{V_{in}} = \frac{(1 + s\tau_1)(1 + s\tau_2) \cdots (1 + s\tau_n)}{(1 + s\tau_a)(1 + s\tau_b) \cdots (1 + s\tau_m)}$$

Points τ_1, τ_2, . . . , τ_n correspond to the zero break frequencies, while points τ_a, τ_b, . . . , τ_m correspond to pole break frequencies. Then plot the gain vs. frequency on logarithmic paper, choosing a gain change slope corresponding to decibels per octave or decibels per decade.

If phase shift is to be plotted, remember that a pole causes a phase lag of $90°$, while a zero causes a phase lead of $90°$ per decade. Since all the information on gain-phase plots is plotted on logarithmic paper in terms of decibels, simple addition of the individual asymptotes is required to derive the final rate of closure.

9-4 FEEDBACK THEORY AND THE CRITERIA FOR STABILITY

Any switching regulator may be treated as a closed-loop feedback control system. A block diagram of a closed-loop system is depicted in Fig. 9-3 where the output signal is fed back and compared to the input. A reference signal $R(s)$ is compared to a feedback signal $B(s)$ at the summing point, and the error signal $E(s)$ inputted to block $G(s)$, and an output $C(s)$ is obtained. In order to derive the closed-loop transfer function $f(s)$, we proceed as follows:

$$C(s) = G(s)E(s)$$
$$B(s) = H(s)C(s)$$
$$E(s) = R(s) - B(s) = R(s) - H(s)C(s)$$

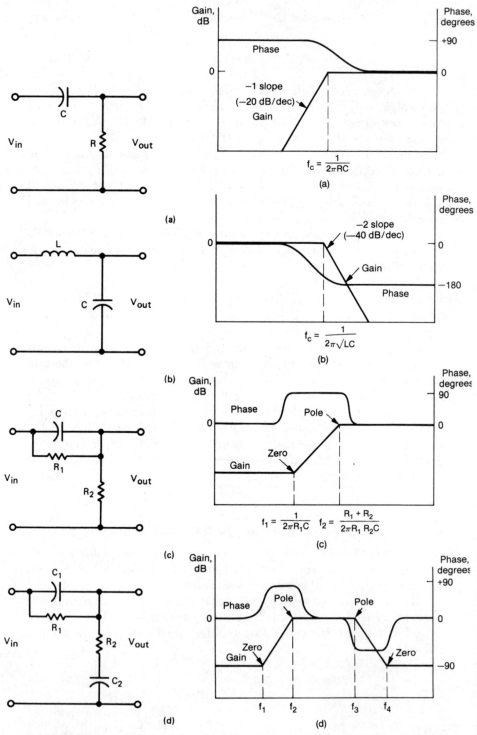

FIGURE 9-2 Networks (*a*) and (*b*) cause poles, while networks (*c*) and (*d*) cause both poles and zeros.

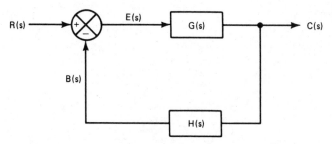

FIGURE 9-3 Block diagram of a closed-loop feedback control system.

Eliminating $E(s)$ from the above equations, we get

$$C(s) = G(s)R(s) - H(s)G(s)C(s)$$

and the closed-loop transfer function is

$$f(s) = \frac{C(s)}{R(s)} = \frac{G(s)}{1 + G(s)H(s)} \tag{9-9}$$

The term $G(s)$ is the open-loop gain, while the term $G(s)H(s)$ is called the open-loop transfer function.

In order to derive a conclusion about the stability of the system, solution of the characteristic equation

$$1 + G(s)H(s) = 0 \tag{9-10}$$

will give the poles of the closed-loop transfer function, since they characterize the response of the system. Therefore, the feedback system must be examined for each value of closed-loop gain to determine the rate of closure between open-loop and closed-loop gain. The objective of stability analysis is to reduce the closed-loop gain roll-off rate to a -1 slope, i.e., -6 dB per octave or -20 dB per decade, at the region of unity gain crossover (0 dB). At that point the phase shift will be less than 360°, a condition for a stable system. The amount the gain is below unity when the total phase shift is 360°, is called the gain margin, while phase margin is the difference between the actual phase shift and 360°, when the loop gain is unity, as shown in Fig. 9-4.

A typical switching power supply closed-loop system is depicted in Figure 9-5. The loop consists of two typical blocks: The modulator, where the power processing takes place, in series with an amplifier generally called the feedback or error amplifier. Although the modulator shown here is a buck reg-

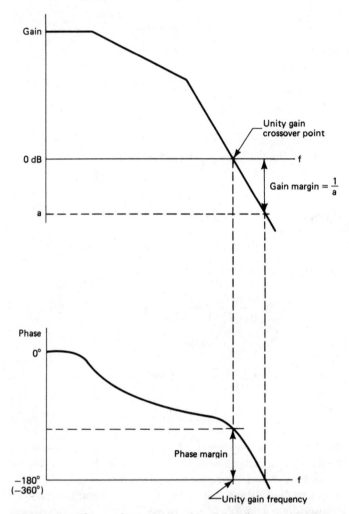

FIGURE 9-4 Phase and gain margin plots of a feedback system. Phase shift is plotted here in terms of 180° because at dc the feedback is negative, i.e., there is an additional 180° of phase shift for a total of 360° phase shift as defined in the text.

ulator type, the discussion to follow holds true for any modulator no matter how complex.

To optimize the design of the feedback loop in order to achieve stable overall operation of the system, the first step is to determine and plot the control-to-output transfer function of the modulator and draw its Bode plot. Typical Bode plots of a modulator showing both gain and phase are shown

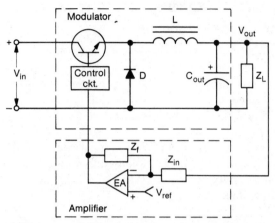

FIGURE 9-5 A typical feedback control loop, showing the
modulator and error amplifier.

in Figs. 9-6 and 9-12. The next step is to determine the unity gain crossover
frequency and the desired phase margin. The unity gain crossover frequency
should be chosen based on desired performance, but a rule of thumb is to
choose a frequency at about one-fifth the regulator's clock frequency.

The desired phase margin, that is, the difference between the actual phase
shift and 360° when the loop gain is unity (0 dB), must be at least 30° or
better. A good compromise is 60°, which also gives a good transient response.
The final step then is to compensate the feedback or error amplifier so that
its gain is equal to the reciprocal of the modulator gain at the desired frequency.

To achieve overall system stability and adequate phase margin, the amplifier gain combined with the modulator gain should produce an overall
gain plot which crosses the unity gain (0 dB) line at the desired crossover
frequency, at a −1 slope, as shown in Fig. 9-14. The −1 slope introduces
90° of phase lag, which combined with the 180° of phase shift already existing
at dc (inverting amplifier), gives a total phase lag of 270°. Since there is 90°
left to get to 360°, the phase margin will be 90°. Remember, as mentioned
in Sec. 9-3, that a −1 slope introduces 90° of phase lag, a −2 slope, 180°
of phase lag, and a −3 slope, 270° of phase lag. It is therefore obvious that
a loop system with a −2 slope at unity gain crossover has no phase margin
because the total phase shift is 360°, while a loop system with a −3 slope
will oscillate because the total phase shift around the loop exceeds 360°.

The following paragraphs of this chapter will examine analysis and design
techniques in tailoring the gain of the error amplifier, allowing prediction
and plotting of loop performance of any switching regulator, without tedious
trial-and-error efforts.

9-5 OFF-THE-LINE SWITCHING POWER SUPPLY STABILITY ANALYSIS

9-5-1 Control-to-Output Transfer Function

All off-the-line PWM switching power supplies consist, more or less, of a modulator, an error amplifier, an isolation transformer, and an output LC filter. The control-to-output transfer function of a switching power supply using a PWM controller IC includes the gain of the sawtooth modulator, the power switching circuit, and the output filter characteristics.

In single-port direct duty cycle control PWM power supply topologies, a voltage V_C applied to the control port of the PWM comparator (see Figs. 7-3 and 7-5) is compared to a sawtooth voltage of constant amplitude V_s to change the comparator's output duty cycle from 0 to 1. The resulting duty cycle δ of the drive waveform then varies as

$$\delta = \frac{V_C}{V_S} \tag{9-11}$$

The gain of the buck family converters, i.e., feed-forward, push-pull, and bridge converters, is given by

$$\frac{V_{out}}{V_{in}} = \frac{N_S}{N_P} \delta = \frac{N_S}{N_P} \frac{V_C}{V_S} \tag{9-12}$$

where N_S/N_P is the transformer secondary-to-primary turns ratio, and V_{in} is the transformer primary voltage.

The gain of the buck-boost converter, i.e., the flyback, is given by

$$\frac{V_{out}}{V_{in}} = \frac{N_S}{N_P} \frac{\delta}{1 - \delta} = \frac{N_S}{N_P} \frac{V_C}{V_S - V_C} \tag{9-13}$$

In order to obtain the control-to-output voltage dc gain of a PWM power supply, Eqs. 9-12 and 9-13 are differentiated with respect to V_C, i.e., $\partial V_{out}/\partial V_C$. For the buck family of converters,

$$\frac{\partial V_{out}}{\partial V_C} = (\text{dc gain}) = \frac{V_{in}}{V_S} \frac{N_S}{N_P} \tag{9-14}$$

In decibels the dc gain is given by

$$(\text{dc gain})_{dB} = 20 \log_{10}\left(\frac{V_{in}}{V_S} \frac{N_S}{N_P}\right) \tag{9-15}$$

For the buck-boost family of converters,

$$\frac{\partial V_{out}}{\partial V_C} = (\text{dc gain}) = \frac{V_{in}}{(V_S - V_C)^2} \frac{N_S}{N_P} = \frac{(V_{in} + V_{out})^2}{V_{in} V_S} \frac{N_S}{N_P} \tag{9-16}$$

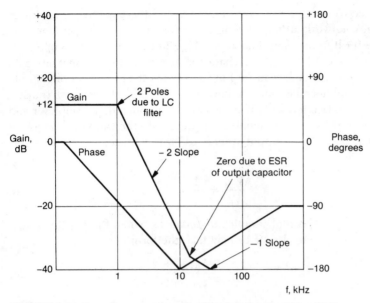

FIGURE 9-6 Control-to-output transfer function characteristics of *LC* filter and modulator, of a PWM switching power supply. The −1 slope at about 20 kHz is caused by a zero due to the ESR of the output filter capacitor. The phase is also shown.

In decibels the dc gain is given by

$$(\text{dc gain})_{\text{dB}} = 20 \log_{10}\left[\frac{(V_{\text{in}} + V_{\text{out}})^2}{V_{\text{in}}V_S}\frac{N_S}{N_P}\right] \tag{9-17}$$

The output filter on the other hand is normally an *LC* type, with a slope of −2 (−40 dB per decade), as shown in Fig. 9-2*b*. The overall closed-loop gain of the power supply is therefore

$$\frac{\partial V_{\text{out}}}{\partial V_C} = [(\text{Gain})\, H(s)] \tag{9-18}$$

This means that the Bode curve of Eq. 9-18 has a dc gain which is flat out to the resonant frequency of the *LC* filter and then falls at a −2 (−40 dB per decade) slope, as shown in Fig. 9-6.

9-5-2 Error Amplifier Compensation

In the majority of the PWM control ICs, the error amplifier is a high-gain operational amplifier, which generates the error signal to the control input of the modulator. Closing the loop in a PWM switching power supply involves the active role of the error amplifier, and the objective is to design the feedback network around the amplifier such that the overall loop gain crosses the 0-dB (unity gain) line at a -1 (-20 dB per decade) slope.

In order to plot the amplifier characteristics in a Bode format, its gain must be written in a Laplace form. Let us examine the properties of a simple operational amplifier first, and see how we will be able to write its transfer function in a Laplace format. Figure 9-7 shows an operational amplifier and its feedback impedances. As mentioned previously the transfer function of this circuit is the ratio of output voltage to input voltage, and for an operational amplifier

$$\frac{V_{out}}{V_{in}} = \frac{Z_f}{Z_i} \tag{9-19}$$

Since Z_f and Z_i represent complex impedances, when the Laplace transform is applied to them, Eq. 9-19 may take the form

$$\frac{V_{out}}{V_{in}} = \frac{(\tau_1 s + 1)(\tau_2 s + 1)}{(\tau_3 s)(\tau_4 s + 1)} \tag{9-20}$$

The operator τ represents a time constant RC in Eq. 9-20. The terms in the numerator represent zeros, while those in the denominator represent poles. The term $(\tau_3 s)$ represents a pole at the origin because it lacks the $+1$ term.

In order to appreciate the ease by which one can write the transfer function of any operational amplifier with complex impedances, and also sketch its transfer function in a Bode plot, the circuit in Fig. 9-8 will be used as an example. By inspection we write,

$$\frac{V_{out}}{V_{in}} = \frac{Z_f}{Z_i} = \frac{R_3 + 1/C_2 s}{R_2 + \{R_1(1/C_1 s)/[R_1 + (1/C_1 s)]\}} \tag{9-21}$$

$$\frac{V_{out}}{V_{in}} = \frac{R_3 + 1/C_2 s}{R_2 + \{(R_1/C_1 s)/[(R_1 C_1 s + 1)/C_1 s]\}} = \frac{R_3 + 1/C_2 s}{R_2 + [R_1/(R_1 C_1 s + 1)]}$$

$$= \frac{(R_3 C_2 s + 1)/C_2 s}{(R_1 R_2 C_1 s + R_1)/(R_1 C_1 s + 1)} = \frac{(R_3 C_2 s + 1)/C_2 s}{R_1(R_2 C_1 s + 1)/(R_1 C_1 s + 1)}$$

$$= \frac{(R_3 C_2 s + 1)(R_1 C_1 s + 1)}{(R_1 C_2 s)(R_2 C_1 s + 1)} \tag{9-22}$$

Comparing Eq. 9-22 to Eq. 9-20,

$$\tau_1 = R_3 C_2$$

$$\tau_2 = R_1 C_1$$

$$\tau_3 = R_1 C_2$$

$$\tau_4 = R_2 C_1$$

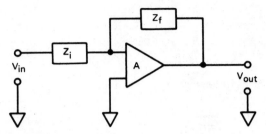

FIGURE 9-7 A simple operational amplifier circuit with feedback impedances.

The high-frequency amplifier gain is set by R_2 and R_3. Therefore the gain at f_3 is

$$AV_2 = \frac{R_3}{R_2} \tag{9-23}$$

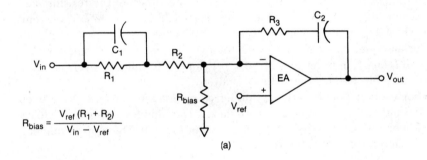

$$R_{bias} = \frac{V_{ref}(R_1 + R_2)}{V_{in} - V_{ref}}$$

(a)

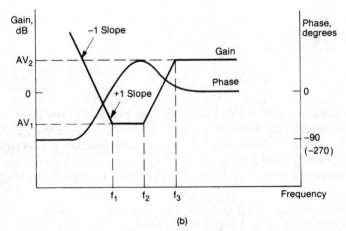

(b)

FIGURE 9-8 (a) Operational amplifier with feedback impedances and (b) its gain Bode plot showing zero-pole pairs. The phase is also shown.

The gain at f_1 and f_2 is

$$AV_1 = \frac{R_3}{R_1 + R_2} \tag{9-24}$$

The Bode plot break or corner frequencies are determined by

$$f_1 = \frac{1}{2\pi R_1 C_1} \tag{9-25}$$

$$f_2 = \frac{1}{2\pi R_3 C_2} \tag{9-26}$$

and

$$f_3 = \frac{1}{2\pi R_2 C_1} \tag{9-27}$$

In an actual design application, the break frequencies are normally predetermined by the design objectives. Then the values of the resistors and capacitors may be easily calculated using Eqs. 9-23 through 9-27. The circuit shown in Fig. 9-8a may be used in any PWM switching power supply for error amplifier compensation, with the overall loop gain crossing over between f_2 and f_3 in order to achieve a -1 slope at 0 gain crossover, which is the final objective of loop stability analysis.

Some other popular error amplifier compensation networks have been published, and they are presented here. The simplest form of feedback amplifier with a single-pole rolloff is shown in Fig. 9-9. The transfer function of the amplifier in Fig. 9-9 is

$$\frac{V_{out}}{V_{in}} = \frac{1}{RCs} \tag{9-28}$$

with a break frequency

$$f_c = \frac{1}{2\pi RC} \tag{9-29}$$

Another amplifier configuration is shown in Fig. 9-10, where a zero-pole pair has been introduced to give a region of frequency where the gain is flat and no phase shift is introduced. The region with constant gain occurs between the break frequencies f_1 and f_2. This region must be used for loop gain crossover, when the circuit is used as an error amplifier in PWM power supplies.

The previously discussed method of analysis may be used in this circuit also to derive the gain and break frequencies, the results of which are as

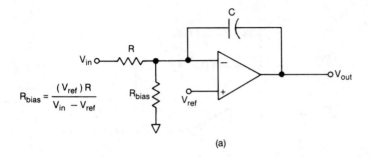

$$R_{bias} = \frac{(V_{ref})R}{V_{in} - V_{ref}}$$

(a)

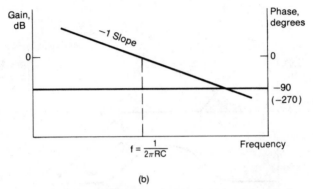

(b)

FIGURE 9-9 (a) A single-pole feedback amplifier and (b) its gain and phase plots. Also known as a type 1 amplifier.

follows

$$AV = \frac{R_2}{R_1} \tag{9-30}$$

$$f_1 = \frac{1}{2\pi R_2 C_1} \tag{9-31}$$

$$f_2 = \frac{C_1 + C_2}{2\pi R_2 C_1 C_2} \approx \frac{1}{2\pi R_2 C_2} \tag{9-32}$$

The amplifiers of Figs. 9-8 and 9-10 offer improved power supply transient response when the supply is subjected to output load changes, as opposed to the slow response of the amplifier shown in Fig. 9-9.

Although it seems to be complicated, the amplifier depicted in Fig. 9-11 will give a very good transient response. In this circuit two zero-pole pairs have been introduced to give a region of frequency where the gain is increased at a $+1$ slope with a $90°$ phase lead. The performance of this amplifier

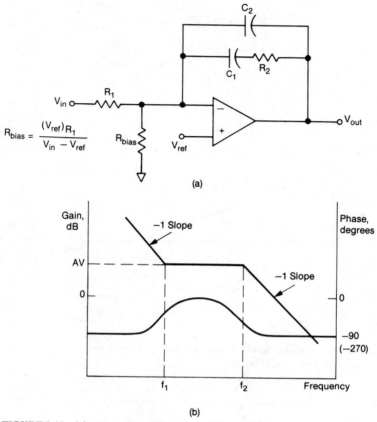

FIGURE 9-10 (*a*) An operational amplifier with a zero-pole pair and (*b*) its gain and phase plot. Also known as a type 2 amplifier.

is very similar to the one shown in Fig. 9-8. The gains and break frequencies are found to be as follows:

$$AV_1 = \frac{R_2}{R_1} \tag{9-33}$$

$$AV_2 = \frac{R_2(R_1 + R_3)}{R_1 R_3} \approx \frac{R_2}{R_3} \tag{9-34}$$

$$f_1 = \frac{1}{2\pi R_2 C_1} \tag{9-35}$$

$$f_2 = \frac{1}{2\pi (R_1 + R_3) C_3} \approx \frac{1}{2\pi R_1 C_3} \tag{9-36}$$

$$f_3 = \frac{1}{2\pi R_3 C_3} \tag{9-37}$$

$$f_4 = \frac{C_1 + C_2}{2\pi R_2 C_1 C_2} \approx \frac{1}{2\pi R_2 C_2} \tag{9-38}$$

When the circuit of Fig. 9-11a is used as a compensated error amplifier in a PWM switching power supply, loop crossover should occur between f_2 and f_3 for better results.

Although many more circuits may be used as error amplifiers with compensation, the four circuits presented here should be adequate to use in the

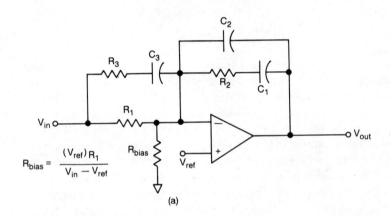

(a)

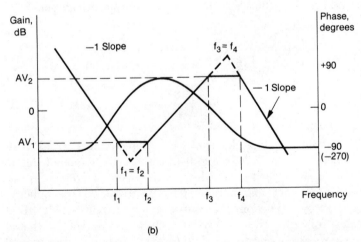

(b)

FIGURE 9-11 (*a*) An operational amplifier with two zero-pole pairs and (*b*) its gain and phase plots. Also known as a type 3 amplifier.

majority of PWM switching power supplies for loop stability analysis and design. The following example shows step-by-step stability analysis in an actual switching application, using theory presented in the above paragraphs.

EXAMPLE 9-2

Consider a half-bridge power supply designed to accept both 90 to 130 V ac or 180 to 260 V ac, working at 20 kHz and using the UC3524A PWM control circuit. The LC output filter has been designed to have a resonant corner frequency at 1 kHz. The power transformer primary-to-secondary turns ratio $N_P/N_S = 18$. Design the error amplifier compensation network in order to achieve overall power supply stability, and sketch the overall loop gain Bode plot.

SOLUTION

First let us choose an amplifier configuration from the four circuits presented in this chapter. Although with careful design all the amplifiers will work, the one in Fig. 9-11 is chosen, since it will give good transient response.

The second consideration is to choose the crossover frequency where the gain is unity and the Bode plot crosses at a -1 slope (-20 dB per decade). Theoretical limits set the crossover frequency at half the switching frequency, but from practical experience less than one-fifth of the switching frequency is used. For this analysis we choose a crossover frequency of 4 kHz, which is one-fifth the switching frequency and one-tenth the modulator frequency.

Since we are using the UC3524A, the control voltage V_C swings 2.5 V to change the comparator's drive waveform from 0 to 1. We also take the worst case of input voltage, i.e., 130 V ac. The control-to-output voltage gain using Eq. 9-15 is

$$(\text{dc gain})_{dB} = 20 \log_{10}\left(\frac{V_{in}}{V_S}\frac{N_S}{N_P}\right) = 20 \log_{10}\left(\frac{182}{2.5}\frac{1}{18}\right)$$

$$= 20 \log_{10}\left(\frac{182}{45}\right) = 20 \log_{10} 4.04 = +12 \text{ dB}$$

The output transfer function characteristic is depicted in Fig. 9-12. Although in practice the asymptote of Fig. 9-12 has a break frequency due to the ESR of the output capacitor, its effect on the overall loop gain in this example is of no importance, and therefore it has been omitted for simplicity.

From Fig. 9-12 by inspection, the control-to-output gain is $+12$ dB at low frequencies, rolling off above 1 kHz at -40 dB per decade, so that at the chosen crossover frequency of 4 kHz the control-to-output gain is -12 dB. The fact that the magnitude of both gains is $|12$ dB$|$ is purely coincidental.

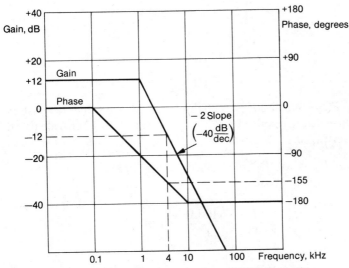

FIGURE 9-12 Control-to-output transfer function of Example 9-1.

Therefore, for an overall loop gain of zero, the feedback amplifier gain must be made +12 dB at 4 kHz.

The important thing to remember here is that the switching power supply will be stable if the overall loop gain crosses over the 0-dB line at a −1 slope. Since the control-to-output gain of the converter is falling at a −2 slope, as depicted in Fig. 9-12, the feedback amplifier must supply a +1 slope at this point for a resulting slope of −1 (−20 dB per decade). Again, the feedback amplifier gain at 4 kHz must be +12 dB (or 4.0) with a +1 slope. Since the input line voltage to the switching power supply swings from low to high line, the +1 slope must have some margin to span the range of crossover frequencies as the modulator's gain changes with input voltage.

Now, let us find the modulator gain at 1 kHz. It will be

$$AV_1 = \frac{1 \text{ kHz}}{4 \text{ kHz}} (4.0) = 1.00 \quad \text{or} \quad 0 \text{ dB}$$

Let us assume, then, the following characteristics for the feedback amplifier and plot its Bode graph. The gain is +12 dB at 4 kHz and 0 dB at 1 kHz. We desire a double zero at 1 kHz, a pole at 10 kHz, and a second pole at 30 kHz. The Bode plot is given in Fig. 9-13. From the graph we get

$$AV_1 = 0 \text{ dB} \quad \text{or} \quad 1.00$$

$$AV_2 = 19.96 \text{ dB} \quad \text{or} \quad 9.95$$

and

$$f_1 = f_2 = 1 \text{ kHz}$$
$$f_3 = 10 \text{ kHz}$$
$$f_4 = 30 \text{ kHz}$$

Referring to Fig. 9-11a and Eqs. 9-33 through 9-38, the values of resistors and capacitors to give the required results depicted in Fig. 9-13 are calculated as follows. From Eq. 9-33, assuming $R_1 = 10 \text{ k}\Omega$,

$$R_2 = AV_1(R_1) = (1.00)(10 \text{ k}\Omega) = 10 \text{ k}\Omega$$

From Eq. 9-34

$$R_3 = \frac{R_2}{AV_2} = \frac{10 \text{ k}\Omega}{9.95} \approx 1 \text{ k}\Omega$$

From Eq. 9-35

$$C_1 = \frac{1}{2\pi f_1 R_2} = \frac{10^{-6}}{62.8} = 0.015 \ \mu F$$

From Eq. 9-36

$$C_3 = \frac{1}{2\pi f_2 R_1} = 0.015 \ \mu F$$

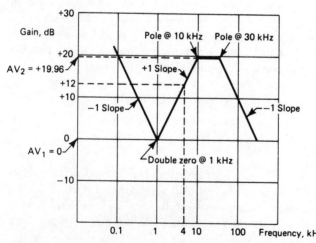

FIGURE 9-13 Feedback amplifier Bode plot showing desired frequency and gain characteristics.

From Eq. 9-38

$$C_2 = \frac{1}{2\pi f_4 R_2} = 0.0005 \ \mu F$$

The final amplifier design and the overall loop gain are shown in Fig. 9-14. Figure 9-14 is plotted by adding the graphs of Figs. 9-12 and 9-13. Notice that the overall gain crosses the 0-dB line (unity gain) at 4 kHz, at a

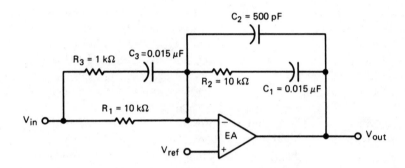

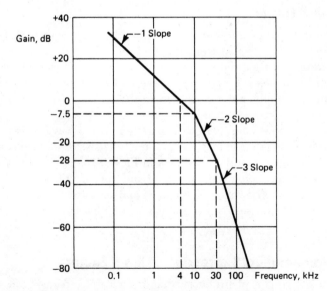

FIGURE 9-14 The compensated feedback amplifier and the overall system loop gain plot of the switching power supply of Example 9-1. (Resistor R_{bias} has no effect and is not shown for simplicity.)

−1 slope, as desired. Of course the crossover frequency will change as the input line voltage is varied over the 90- to 130-V ac (or 130- to 260-V ac) range, but the crossover will still be at a −1 slope. The reader may verify this by plotting the loop gain Bode curves at low line, i.e., 90 V ac (180 V ac).

9-6 STABILITY ANALYSIS AND SYNTHESIS USING THE *K* FACTOR

In the previous paragraphs the subject of stability analysis and synthesis was discussed, and practical mathematical tools were developed to design compensated error amplifiers to be used with any type of switching regulator. Three basic amplifier types were presented and analyzed, and the design equations given with each type have universal usage (see Figs. 9-9 to 9-11).

The following paragraphs will present a new powerful mathematical tool, known as the *K* factor, which makes the analysis and synthesis of error amplifiers even easier. The techniques presented here are valid with any type of modulator used, whether buck, boost, or buck-boost.

It is important to note that in all cases a real operational amplifier is used, either external or part of the IC control circuit, and that the system is configured so that negative feedback is required in the amplifier.

9-6-1 The *K* Factor

The *K* factor is a mathematical tool for defining the shape and characteristics of a transfer function. No matter what type of amplifier is chosen, the *K* factor is a measure of the reduction of gain at low frequencies and increase gain at high frequencies, by controlling the location of the poles and zeros of the feedback amplifier Bode plot, in relationship to the loop crossover frequency *f*.

Figure 9-15*a* shows that for type 1 amplifiers *K* is always 1. This is due to a total lack of phase boost or corresponding increase or decrease in gain.

For amplifier types 2 and 3, as shown in Figs. 9-15*b* and 9-15*c*, the zero is placed a factor of *K* below loop crossover and the pole a factor of *K* above. Since *f* is the geometric mean of the zero and pole locations, peak phase boost will occur at the crossover frequency. For either case, as *K* is increasing, so is the phase boost.

9-6-2 Mathematical Expression of the *K* Factor

It is widely known that phase boost, due to a zero-pole pair, is the inverse tangent ratio of the measurement frequency to the zero or pole frequency. The total phase shift then is the sum of all individual zero and pole phase shifts.

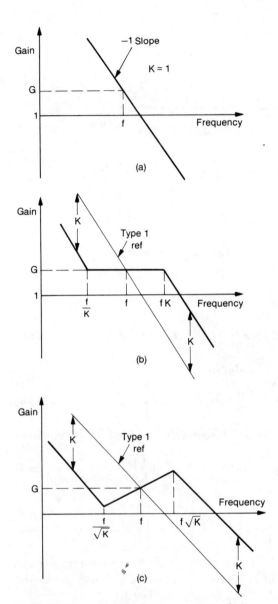

FIGURE 9-15 The Bode plot characteristics of (a) type 1 amplifier, (b) type 2 amplifier, and (c) type 3 amplifier, in relation to the K factor. All coordinates are logarithmic. (See also Figs. 9-9 to 9-11.)

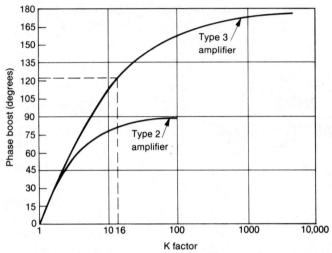

FIGURE 9-16 Phase boost vs. *K* factor for type 2 and type 3 feedback amplifiers.

For type 2 amplifiers the boost at frequency f is given by the equation

$$\text{Boost} = \tan^{-1}(K) - \tan^{-1}(K) \tan^{-1}\left(\frac{1}{K}\right) \qquad (9\text{-}39)$$

and from this equation it can be shown that,

$$K = \tan\left[\left(\frac{\text{boost}}{2}\right) + 45\right] \qquad (9\text{-}40)$$

For type 3 amplifiers the boost at frequency f is given by the equation

$$\text{Boost} = \tan^{-1}K - \tan^{-1}\left(\frac{1}{K}\right) \qquad (9\text{-}41)$$

and subsequently,

$$K = \left\{\tan\left[\left(\frac{\text{boost}}{4}\right) + 45\right]\right\}^2 \qquad (9\text{-}42)$$

The above equations yield the phase boost vs. *K*-factor curves depicted graphically in Fig. 9-16. These curves are universal and may be used to readily determine the *K* factor for a given phase boost.

9-6-3 Synthesis of Feedback Amplifiers Utilizing the *K* Factor

Using the *K* factor to synthesize an amplifier, the following steps are recommended.

Step 1:   *Make a Bode plot of the modulator.* A typical modulator Bode plot showing both gain and phase characteristics is depicted in Fig. 9-12.

Step 2:   *Choose a crossover frequency.* The crossover frequency is the point where you want the overall loop gain to be unity. Remember that the higher the crossover frequency, the better the transient response of the power supply. However, practical limitations restrict the range of the crossover frequency. The theoretical limit is half the switching frequency, but practical considerations have proven that a crossover frequency figure of less than one-fifth of the clock frequency is a good choice.

Step 3:   *Choose the desired phase margin.* This margin is the amount of phase desired at unity gain, as shown in Fig. 9-4 and 9-14. Phase margin may have a range of 30 to 90°, with 60° being a good compromise.

Step 4:   *Determine required amplifier gain.* This gain G is the required amplifier gain at crossover and must equal the modulator loss. When expressed in decibels, the amplifier gain is simply the negative of the modulator gain; otherwise, amplifier gain = 1/modulator gain.

Step 5:   *Calculate the required phase boost.* The amount of phase boost required from the zero-pole pair in the amplifier is given by the formula

$$\text{Boost} = M - P - 90° \qquad (9\text{-}43)$$

where M = desired phase margin, degrees
P = modulator phase shift, degrees

Step 6:   *Choose the amplifier type.* Choose amplifier type 1 when no boost is required, amplifier type 2 when the required boost is less than 90°, and amplifier type 3 when the required phase boost is less than 180°.

Step 7:   *Calculate the K factor.* The K factor may be calculated using either Eq. 9-40 or 9-42 or directly from the curves of Fig. 9-16. For amplifier type 1, $K = 1$. Location of the transfer function poles and zeros will determine the circuit values. The pole at the origin causes the initial -1 gain slope, and the frequency where this line crosses or would have crossed the 0-dB line (unity gain), is the unity gain frequency, UGF.

The following equations provide the component values of each amplifier type.

Type 1
(Figs. 9-9, 9-15a):

$$C = \frac{1}{2\pi fGR} \tag{9-44}$$

Type 2
(Figs. 9-10, 9-15b):

$$UGF = \frac{1}{2\pi R_1(C_1 + C_2)} \tag{9-45}$$

$$C_2 = \frac{1}{2\pi fGKR_1} \tag{9-46}$$

$$C_1 = C_2(K^2 - 1) \tag{9-47}$$

$$R_2 = \frac{K}{2\pi fC_1} \tag{9-48}$$

Type 3
(Figs. 9-11, 9-15c):

$$UGF = \frac{1}{2\pi R_1(C_1 + C_2)} \tag{9-49}$$

$$C_2 = \frac{1}{2\pi fGR_1} \tag{9-50}$$

$$C_1 = C_2(K - 1) \tag{9-51}$$

$$R_2 = \frac{\sqrt{K}}{2\pi fC_1} \tag{9-52}$$

$$R_3 = \frac{R_1}{K - 1} \tag{9-53}$$

$$C_3 = \frac{1}{2\pi f\sqrt{K}R_3} \tag{9-54}$$

EXAMPLE 9-3

Consider the power supply requirements given in Example 9-2. Design the error amplifier compensation network to achieve overall loop stability, with a phase margin of 60° at the unity crossover frequency.

SOLUTION

From the modulator Bode plot depicted in Fig. 9-12, at a crossover frequency of 4 kHz (one-tenth of the modulator frequency and one-fifth of the clock frequency), the amplifier gain must be 4 or −12 dB. By inspection of the same plot, the phase curve and the crossover frequency line show a modulator phase shift of 155°.

Using Eq. 9-43, we calculate the required phase boost:

$$\text{Boost} = 60 - (-155) - 90 = 125°$$

Using Eq. 9-42, or reading directly from the curves in Fig. 9-16, to achieve 125° of phase boost, we need $K = 16$. Now the component values of the feedback network of the amplifier shown in Fig. 9-18 may be calculated using Eqs. 9-50 through 9-54.

To facilitate the calculations, resistor R_1 is arbitrarily chosen to be 10 kΩ. Hence,

$$C_2 = \frac{1}{2\pi f G R_1} = \frac{1}{6.28(4 \times 10^3)4 \times 10 \times 10^3} = 0.001 \ \mu F$$

$$C_1 = C_2(K - 1) = 0.001(16 - 1) = 0.015 \ \mu F$$

$$R_2 = \frac{K}{2\pi f C_1} = \frac{\sqrt{16}}{6.28(4 \times 10^3)(0.001)10^{-6}} = 10 \ k\Omega$$

$$R_3 = \frac{R_1}{K - 1} = \frac{10}{16 - 1} = 670 \ \Omega$$

$$C_3 = \frac{1}{2\pi f \sqrt{K} R_3} = \frac{1}{6.28(4 \times 10^2)\sqrt{16}(670)} = 0.015 \ \mu F$$

Since the crossover frequency f is known to be 4 kHz, by inspection of the curve shown in Fig. 9-15b we note that a double zero is located at a frequency $f/\sqrt{K}$ below the crossover frequency, while a double pole is located at a frequency of $f\sqrt{K}$ above.

The double-zero location is

$$f_z = \frac{f}{\sqrt{K}} = \frac{4}{\sqrt{16}} = 1 \ kHz$$

The double-pole location is

$$f_p = f\sqrt{K} = 4\sqrt{16} = 16 \ kHz$$

For a more accurate location of zeros and poles, Eqs. 9-35 through 9-38 could have been used, but for all practical purposes the above results are acceptable. The amplifier Bode plot is shown in Fig. 9-17.

Unity gain crossover occurs at 1 kHz (verify, using Eq. 9-49). Note that the results of this example are the same as the ones previously calculated in Example 9-2.

The only difference is the values of C_2 and R_3, which resulted in making the poles previously located at 10 and 30 kHz coincide as a double pole located at 16 kHz. This is a direct result of the K-factor assumptions, of a type 3 amplifier, the zeros and poles are coincident. Of course the feedback component values may be adjusted using Eqs. 9-35 through 9-38, so that zeros or poles or both are spread apart. The net effect of this is to broaden and flatten the phase bump, thus reducing the phase margin at crossover.

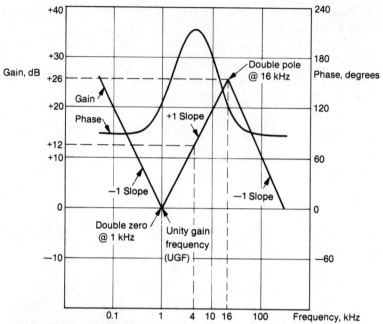

FIGURE 9-17   Amplifier gain Bode plot and phase plot of Example 9-3.

Optimum performance therefore is achieved when both zeros and poles are coincident, although it is strongly suggested that each switching regulator circuit is analyzed for its own merits and accordingly tailor the amplifier design to meet optimum performance.

We have therefore successfully completed the design of the feedback amplifier to produce the desired overall system loop gain and phase plot shown in Fig. 9-18. Hence, the overall gain plot crosses the 0-dB line (unity gain) at 4 kHz, at a −1 slope up to 16 kHz, at which point it falls off at a −3 slope. At the same time the phase plot shows a 60° phase margin at the unity gain crossover, as desired, assuring a stable system operation.

9-7   LOOP STABILITY MEASUREMENTS

Although there are numerous methods of measuring the overall loop gain of a switching power supply (see references), a very simple and useful way of deriving information on closed-loop stability is by measuring the power supply's transient response. Transient response is measured by switching the output load from 75 to 100 percent of its full value at a rate of twice the ac line input frequency. Such a load change forces the feedback amplifier from an open-loop to a closed-loop condition at the end of the recovery time.

Figure 9-19 shows typical transient response traces for a ±25 percent load change. The switching waveform of Figure 9-19a causes the output voltage of the switcher to "dip" or "jump" at the square wave rise and fall edges. The magnitude V_r of these transients depends primarily on the ESR

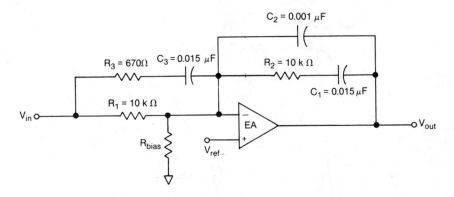

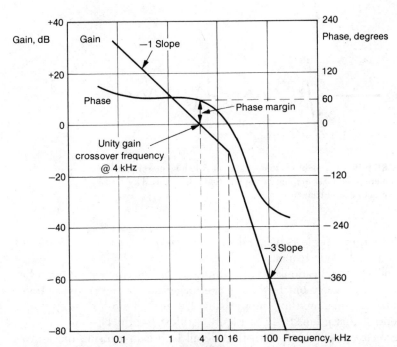

FIGURE 9-18 Overall loop gain and phase plot for the compensated error amplifier of Example 9-3.

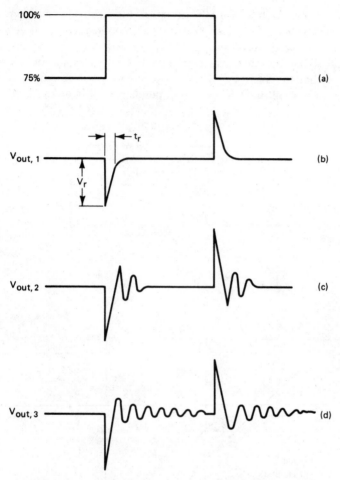

FIGURE 9-19 Transient response traces of switching power supplies with different feedback amplifier compensation values caused by a ±25 percent output load change.

of the output capacitors, while the recovery time t_R is a function of the output filter and the loop response.

For most applications, what is important is not how long it takes for the output to recover, but how large the magnitude of the recovery transient is. For example, anything more than ±250 mV on a 5-V dc output may be potentially hazardous to TTL. Figure 9-19b shows the best and more desirable recovery response, with the overall loop gain crossing the unity gain line at −20 dB per decade and phase margin greater than 90°. Figure 9-19c is also an acceptable recovery response, since the ringing is damped out in a cycle or two. The overall loop gain, in this case, crosses the unity gain line at a slope very close to -20 dB per decade, with a phase margin between 90 and 45°. Figure 9-19d, however, shows marginal stability and tendency of the power supply toward oscillation with very poor phase margin.

Practical loop stability measurement may be accomplished either by manual or computerized methods. Manual methods involve the use of specialized instruments, such as network analyzers, signal generators, etc. More reliable and faster results may be obtained by computerized setups. Venable Industries, Inc., Torrance, California, has introduced a new generation of design tools for frequency response analysis, synthesis, modeling, and testing of power supplies. An engineer can now arrive at an optimized error amplifier or compensator design and documentation in minutes using a personal computer and support hardware. These design tools follow the ideas presented in this chapter and are the result of the exceptional work and contributions which H. Dean Venable, the founder of the company, has made toward the subject of stability analysis. For more information on the products, the reader is encouraged to contact the company directly (see references below).

REFERENCES

1. Bogart, T. F., Jr.: "Laplace Transforms and Control Systems Theory for Technology," Wiley, New York, 1982.

2. Chambers, D., and Wang, D.: Assessing Switcher Performance Call for Fresh Ways and Means, *EDN*, April 1978.

3. Hewlett-Packard: Loop Phase-Gain Measurement of a Power MOSFET Switch Mode Power Supply, *Application Bulletin*, no. 33, 1980.

4. Middlebrook, R. D., and S. Ćuk: "Advances in Switched Mode Power Conversion," vols. 1 and 2, Teslaco, Pasadena, Calif., 1981.

5. Pressman, A. I.: "Switching and Linear Power Supply Power Converter Design," Hayden, Rochelle Park, N.J., 1977.

6. Venable, H. D., and S. R. Foster: Practical Techniques for Analyzing, Measuring and Stabilizing Feedback Loops in Switching Regulators and Converters, *Powercon* 7, 1980.

7. Venable, H. D.: "The K-Factor: "A New Mathematical Tool for Stability Analysis and Synthesis," Powercon 10 Proceedings, March 1983.

8. ———: "Stability Analysis Made Simple," Venable Industries, Inc., Torrance, Calif., 1982.

9. ———: "A New, Easy-to-Use Method for the Design and Analysis of Feedback Control Loops," Venable Industries, Inc., 3555 Lomita Boulevard, Torrance, Calif., 90505.

10. ———: K-Factor Simplifies Linear Amplifier Design, *PCIM*, October 1987.

11. Wood, J. R.: Using the Circle Criterion in the Design and Analysis of Nonlinear Feedback Systems, *Powercon* 9, 1982.

TEN

ELECTROMAGNETIC AND RADIO FREQUENCY INTERFERENCES (EMI-RFI) CONSIDERATIONS

10-0 INTRODUCTION

United States and international standards for EMI-RFI have been established which require the manufacturers of electronic equipment to minimize the radiated and conducted interference of their equipment to acceptable levels. In the United States the guiding document is the FCC Docket 20780, while internationally the West German Verband Deutscher Elektronotechniker (VDE) safety standards have been widely accepted.

It is very important to understand that both the FCC and VDE standards exclude subassemblies from compliance to the rules; rather, the final equipment, where the switching power supply is used, must comply with the EMI-RFI specifications. Rightly so, since even if the switching power supply has an input filter, this filter is matched to the power supply when passive loads are powered, and its characteristics and suppression capabilities may drastically change when used to power active electronic circuits.

This chapter attempts to introduce the reader to the conducted RFI problem and gives some suggestions for minimizing it, whether it is applied in a power supply or a final system.

10-1 THE FCC AND VDE CONDUCTED NOISE SPECIFICATIONS

Both the FCC and VDE are concerned with RFI suppression generated by equipment connected to the ac mains employing high-frequency digital circuitry. The VDE has subdivided its RFI regulations into two categories, the first being unintentional high-frequency generation by equipment with rated frequencies from 0 to 10 kHz, i.e., VDE-0875 and VDE-0879, and the second dealing with intentional high-frequency generation by equipment using frequencies above 10 kHz, i.e., VDE-0871 and VDE-0872.

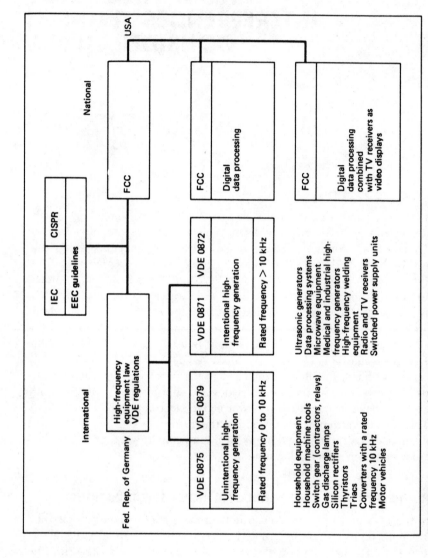

FIGURE 10-1 Summary of the FCC and VDE EMI-RFI requirements.

The FCC on the other hand includes in its RFI regulations all electronic devices and systems which generate and use timing signals or pulses at a rate greater than 10 kHz. Figure 10-1 summarizes the FCC and VDE RFI requirements.

The FCC EMI-RFI regulations closely follow those of the VDE. The FCC class A specification covers business, commercial, and industrial environments, and compliance to the specified EMI emissions in decibel-μvolts can be met by any equipment meeting VDE-0875/N or VDE-0871/A,C.

On the other hand, FCC class B requirements cover residential environments and are more stringent than those of class A. Both FCC conducted EMI-RFI specifications, however, cover the frequency range from 450 kHz to 30 MHz. The VDE regulations extend below the 450-kHz range; in fact the VDE frequency range for EMI-RFI conducted emissions covers a spectrum from 10 kHz to 30 MHz. Figure 10-2 shows the FCC and VDE curves for conducted RFI emissions.

10-2 RFI SOURCES IN SWITCHING POWER SUPPLIES

Every switching power supply is a source of RFI generation because of the very fast rise and fall times of the current and voltage waveforms inherent

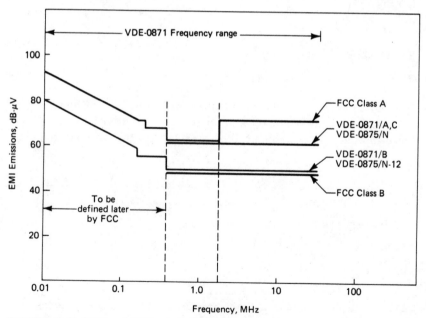

FIGURE 10-2 FCC and VDE regulation curves showing maximum permissible RFI emissions in decibels-microvolt on conducted noise.

in the converter operation. The main sources of switching noise are the switching transistor(s), the mains rectifier, the output diodes, the protective diode for the transistor, and of course the control unit itself. Depending upon the topology of the converter used, the RFI noise level at the mains input may vary from bad to worse.

Flyback converters, which by design have a triangular input current waveform, generate less conducted RFI noise than converters with rectangular input current waveforms, such as feed-forward or bridge converters. Fourier analysis shows that the amplitudes of the high-frequency harmonics of a triangular current waveform drop at a rate of 40 decibels per decade, compared to a 20 decibel per decade drop for a comparable rectangular current waveform.

10-3 AC INPUT LINE FILTERS FOR RFI SUPPRESSION

The most common method of noise suppression at switching power supply ac mains is the utilization of an LC filter for differential- and common-mode RFI suppression. Normally a coupled inductor is inserted in series with each ac line, while capacitors are placed between lines (called X capacitors) and between each line and the ground conductor (called Y capacitors).

The capacitance and inductance of the components may be within the following values:

C_X: 0.1 to 2 μF

C_Y: 2200 pF to 0.033 μF

L: 1.8 mH at 25 A to 47 mH at 0.3 A

Figure 10-3 depicts a standard switching power supply input line filter.

During filter component selection it is important to make sure that the resonant frequency of the input filter is lower than the working frequency

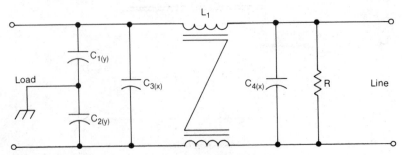

FIGURE 10-3 A switching power supply input line filter for ac mains RFI noise suppression.

of the power supply. On the other hand, filtering conducted noise becomes much easier as the working frequency of the power supply is increased.

The resistor R across the ac lines of the filter is a discharge resistor for the X capacitors, and it is recommended by the safety specifications of the VDE-0806 and IEC-380. In fact IEC-380 Sec. 8.8, states that if the RFI X capacitor is above 0.1 μF, a discharge resistor of the following value is required in the circuit:

$$R = \frac{t}{2.21C} \tag{10-1}$$

where $t = 1$ s, and C is the sum of the X capacitors (in microfarads).

EXAMPLE 10-1

Calculate the discharge resistor R for the filter of Fig. 10-3, given that $C3(X) = C4(X) = 0.1 \mu$F.

SOLUTION

Using Eq. 10-1

$$R = \frac{t}{2.21C} = \frac{1}{(2.21)(0.2)} = 2.2 \text{ M}\Omega$$

Further reduction of the symmetrical and asymmetrical interference voltage may be accomplished by the insertion of an extra line choke, $L2$, as shown in Fig. 10-4. Insertion of choke L_2 leads to a limitation of the charging current of capacitor $C4(X)$.

Although the described circuits will suppress the generated RFI to acceptable levels, it is important to understand that if the power supply packaging or layout changes, a certain filter may or may not work properly. To elaborate on this claim, if a power transistor or power rectifier which uses

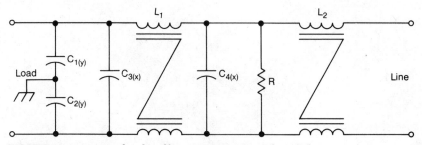

FIGURE 10-4 Improved ac line filter incorporating two line chokes.

high-frequency waveforms is directly mounted on the chassis of the power supply, with only a mica insulator between the two, and if the chassis is connected to the ac ground conductor, generated RF noise will be coupled into the ground conductor, thus upsetting the effectiveness of the particular mains filter. It has been shown that a TO-3 switching transistor working at 20 kHz with a 200-V input, mounted on a ground heat sink through a mica insulator, will generate an RF current of 1 mA at 1 MHz. A solution is to sandwich a metal shield between the insulators and to return the shield to the dc ground. This technique effectively "shorts out" the capacitor created by the mica insulator, reducing RF noise currents.

Power supply and system layout are very important in reducing or eliminating RFI-EMI problems. The designer should take care in analyzing all potential problems before the proper line filter is chosen.

REFERENCES

For complete information on EMI-RFI regulations, the following original documents are recommended.

1. VDE-0875/6.77   "Specification for the Radio Interference Suppression of Equipment, Machines, and Systems with Operational Frequencies from 0 to 10 kHz."

2. VDE-0871/6.78   "Regulations for the Radio Frequency Interference Suppression of High Frequency Apparatus and Installations."

3. FCC Docket 20780   "First Report and Order for Technical Standards for Computing Equipment," part 15, subpart J.

4. Docket 80-284, FCC 81-69   "FCC Methods of Measurement of Radio Noise Emissions from Computing Devices."

ELEVEN

POWER SUPPLY ELECTRICAL SAFETY STANDARDS

11-0 INTRODUCTION

National and international safety regulatory agencies have established electrical safety standards, formulated and directed toward the manufacturing of equipment and/or electrical components, to provide the end user with a safe, quality product. These standards aim to prevent injury or damage due to electrical shock, fire, mechanical, and heat hazards, etc.

In general, each country may impose local standards for electrical safety, but most power supply manufacturers use the IEC (International Electrotechnical Commission), VDE, UL (Underwriters' Laboratories) and CSA (Canadian Standards Association) standards as a de facto solution to the majority of the world's safety requirements. The West German safety standard for business machines, VDE-0806, is based on the IEC's recommendation IEC-380, and it is by far the most stringent electrical safety standard for power supplies. For the United States and Canada, power supplies are generally designed to meet safety standards for data processing equipment, i.e., UL-478 and CSA C22.2 no. 154-1975, and office equipment safety standards, i.e., UL-144 and CSA-C22.2 no. 143-1975.

In this book VDE, UL, and CSA safety standards refer to the above requirements, unless otherwise specified.

11-1 POWER SUPPLY CONSTRUCTION REQUIREMENTS FOR SAFETY

11-1-1 Spacing Requirements

The UL, CSA, and VDE safety specifications impose specific spacing requirements for between live parts and between live and dead metal parts. The UL and CSA require that high-voltage conductors of opposite polarity up to 250 V ac, or high-voltage conductors and dead metal parts, other than field wiring terminals, must have a separation distance of 0.10 in., either over surface or through air. The VDE requires a 3-mm creepage or a 2.0-mm clearance distance between ac lines, and a 4-mm creepage or a 3-mm clearance distance between ac lines and the grounding conductor. The IEC has even tougher specifications, requiring a 3-mm clearance distance between ac lines and a 4-mm clearance distance between ac lines and the grounding conductor. In addition the VDE and IEC require a full 8-mm spacing between the input and output sections of the power supply. Notice that what the UL calls separation over surface, the VDE calls creepage, while the UL definition of separation through air corresponds to the VDE clearance distance.

Figure 11-1 shows the distinction between the measurement of clearance and creepage distances. In Fig. 11-1a the path under consideration includes a V-shaped groove with an internal angle of less than 80° and with a width greater than 1 mm, as well as a parallel or converging-sided groove of any depth with width less than 1 mm. The rule in this case states that clearance is "line-of-sight" distance, and it is measured over the grooves. The creepage distance is measured on the surface of the grooves, as shown, but short-circuits the bottom of the V-shaped groove by 1 mm. The contribution to the creepage of any groove less than 1 mm wide is limited to its width; that is, only the clearance distance applies. Figure 11-1b shows a path which includes a rib. In this case, clearance is the shortest direct air path over the top of the rib, while the creepage path follows the contour of the rib.

Figure 11-2 shows different examples of printed circuit board design to achieve clearance and creepage distances between primary and secondary circuits of a power supply. As shown in Fig. 11-2a, if the primary circuit track is opposite to the secondary circuit track, the thickness of the printed circuit board must be 2 mm minimum. When the printed circuit board is greater than 1 mm but less than 2 mm, then the primary and secondary circuit tracks must be separated by at least 3 mm, as shown in Fig. 11-2b. If the primary and secondary circuit tracks face each other, as shown in Fig. 11-2c, then the full 8-mm clearance distance applies.

11-1-2 Dielectric Test Withstand

For equipment rated 250 V ac or less, both UL and CSA specifications require an input-to-output and input-to-ground hi-pot isolation test of 1000 V ac for 1 min, or 1200 V ac for 1 s. This ac potential must be a sine wave of 50 or 60 Hz.

The VDE requires the following dielectric tests: 3750 V ac between each input ac line and secondary extra low voltage (SELV) output circuits; 2500 V ac between ac lines and the grounding conductor; 500 V ac between the

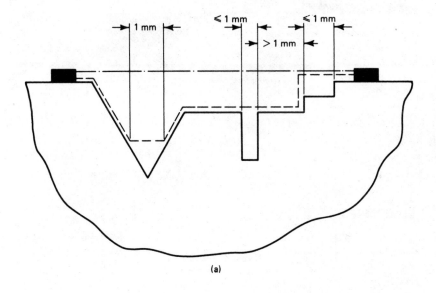

(a)

Legend: _ _ _ Creepage distance
_ . _ Clearance distance

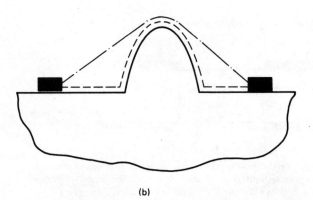

(b)

FIGURE 11-1 Measurement of clearance and creepage distances as specified in the VDE safety standards. (a) Path with a V-shaped groove, (b) path including a rib.

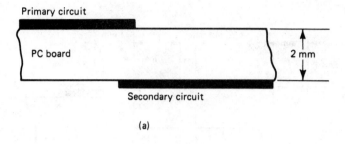

(a)

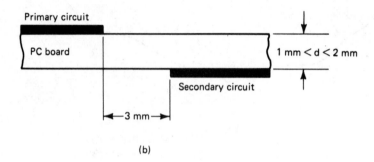

(b)

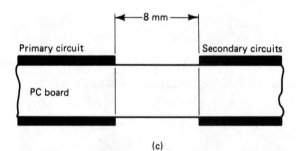

(c)

FIGURE 11-2   Proper printed circuit board design to meet VDE clearance and creepage distance requirements between primary and secondary circuits of a power supply. (a) With primary circuit track opposite secondary circuit track; (b) with printed circuit board greater than 1 mm but less than 2 mm; (c) with primary and secondary circuits facing each other.

grounding conductor and secondary SELV output circuits; and 1250 V ac between ac input lines. All tests are 1 min in duration. The test may be reduced to 1 s if all test voltages are increased by 10 percent.

11-1-3 Leakage Current Measurements

The UL and CSA require that all exposed dead metal parts must be earth-grounded and that the leakage current measured through a 1500-Ω resistor connected to earth ground must not exceed 5 mA.

The VDE allows the following leakage values, measured at 1.06 times rated voltage through a 1500-Ω resistor in parallel with a 150-nF capacitor: for portable office equipment (<25 kg), 0.5 mA; for nonportable office equipment, 3.5 mA; and for data processing equipment, 3.5 mA, maximum.

It is interesting to note that Japan allows a maximum leakage current of 1 mA, measured through a 1000-Ω resistor, for line frequencies up to 1 kHz. For higher leakage currents an isolation transformer at the installation is required. For line frequencies above 1 kHz the maximum leakage current is logarithmically increasing to a value of 20 mA at 30 kHz.

11-1-4 Insulation Resistance

VDE requires 7.0 MΩ minimum resistance between input and SELV output circuits, and 2.0 MΩ between input and accessible metal parts, with an applied voltage of 500 V dc for 1 min.

11-1-5 PC Board Requirements

The UL and CSA flammability standards apply, i.e., all pc boards must be UL-recognized 94V-2 or better material. The VDE accepts these standards.

11-2 POWER SUPPLY TRANSFORMER CONSTRUCTION FOR SAFETY

Since the VDE standards for the design, manufacture, and utilization of transformers impose the most stringent specifications satisfying the majority of safety requirements of other countries, they will be presented here in depth. Because the VDE has no flammability requirements for transformer construction, the UL standards may be used, which require that all material used in the construction of transformers must have a 94V-2 or better rating.

11-2-1 Transformer Insulation

The windings of a transformer must be separated physically by insulation in accordance with the requirements shown in Fig. 11-3 and Table 11-1.

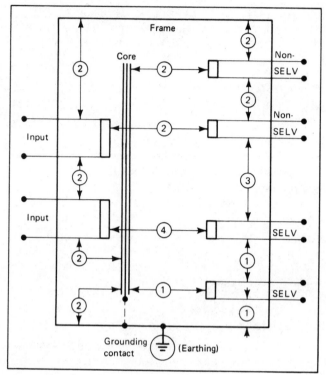

FIGURE 11-3 Distances through transformer insulation as specified by the VDE.

Enamel, lacquer, or varnish coatings on winding wire or other metallic parts, asbestos, and moisture-absorbing material are not considered to be insulation within the meaning of this requirement.

11-2-2 Transformer Dielectric Strength

When multiple layers of plies of insulation are employed, any two layers must be capable of withstanding the dielectric strength value shown in Fig. 11-4, where the layers are in contact and where the test potential is applied to the outer surfaces. The applied ac potential must be a sine wave of 50 or 60 Hz, and the test duration must be 1 min. No insulation rupture or flashover may occur during the dielectric strength test.

11-2-3 Transformer Insulation Resistance

The insulation employed in the construction of a transformer must possess a minimum resistance of 10 MΩ between windings, and between windings

TABLE 11.1   DISTANCE THROUGH TRANSFORMER INSULATION

Ref. no	Working voltage		
	$U < 50*$	$50 \leq U \leq 250$	$U > 250$
1	1 ply†	—	—
2	1 ply	2 plies or 0.5	2 plies or 0.8
3	3 plies or 0.5 or 2 plies/screen/2 plies	3 plies or 0.5 or 2 plies/screen/2 plies	3 plies or 0.8
4	—	3 plies or 2.0 or 2 plies/screen/2 plies	—

*The symbol U signifies the working voltage between the indicated points.

†The minimum thickness for plies is 0.1 mm.

and core and frame and screen, with an applied voltage of 500 V dc for 1 min.

11-2-4   Transformer Creepage and Clearance Distances

The spacings between windings; between windings and terminals, screen, core, frame, winding crossover leads; between terminals; and between terminals—core and frame—must be in accordance with the values shown in Fig. 11-5 and Table 11-2. Creepage and clearance values are based on the assumption that the winding wire is coated with varnish and the like.

11-2-5   Transformer Moisture Resistance

The transformer must be capable of complying with insulation resistance requirements and dielectric strength requirements immediately after the transformer has been subjected to adverse humidity conditions, where the relative humidity is 92 ± 2 percent and the stabilized temperature value between 20 and 30°C with a stabilization factor of ±1°C. The duration of conditioning is 48 h, minimum. The transformer may be temperature-stabilized not more than 4°C greater than the humidity conditioning temperature value prior to conditioning.

11-2-6   VDE Transformer Temperature Rating

The maximum stabilized temperature under normal operation for a specific insulation class must not exceed the temperature value of the insulation classes as shown in the table below. Consideration must be given to the utilization ambient temperature within the product or power supply during temperature evaluation.

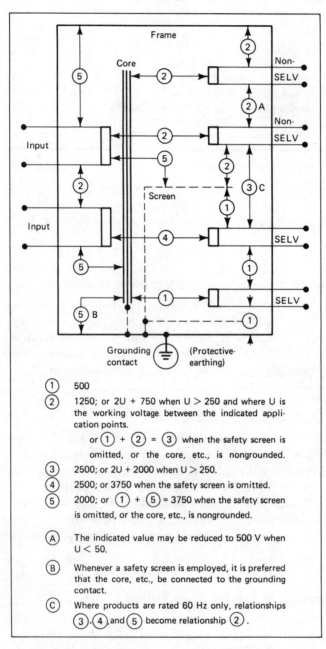

① 500

② 1250; or 2U + 750 when U > 250 and where U is the working voltage between the indicated application points.

or ① + ② = ③ when the safety screen is omitted, or the core, etc., is nongrounded.

③ 2500; or 2U + 2000 when U > 250.

④ 2500; or 3750 when the safety screen is omitted.

⑤ 2000; or ① + ⑤ = 3750 when the safety screen is omitted, or the core, etc., is nongrounded.

Ⓐ The indicated value may be reduced to 500 V when U < 50.

Ⓑ Whenever a safety screen is employed, it is preferred that the core, etc., be connected to the grounding contact.

Ⓒ Where products are rated 60 Hz only, relationships ③, ④, and ⑤ become relationship ②.

FIGURE 11-4 VDE transformer dielectric strength.

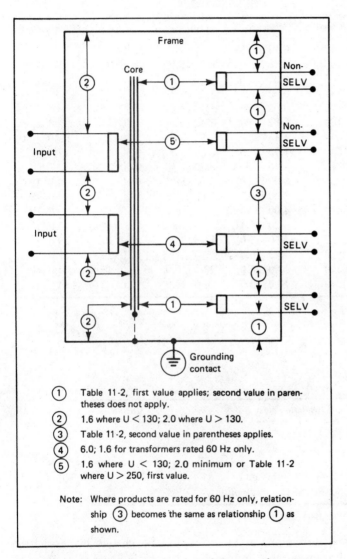

①   Table 11-2, first value applies; **second value in paren-theses does not apply.**

②   1.6 where U < 130; 2.0 where U > 130.

③   Table 11-2, second value in parentheses applies.

④   6.0; 1.6 for transformers rated 60 Hz only.

⑤   1.6 where U < 130; 2.0 minimum or Table 11-2 where U > 250, first value.

Note: Where products are rated for 60 Hz only, relation-ship ③ becomes the same as relationship ① as shown.

FIGURE 11-5   Transformer creepage and clearance distances.

TABLE 11-2 CREEPAGE AND CLEARANCE REQUIREMENTS IN SECONDARY CIRCUITS (VALUES IN MILLIMETERS)

U = Working voltage		50 Hz all VA 50/60, 50–60 Hz ≤ 200 VA		>200 VA 50/60 or 50–60 Hz, 60 Hz	
Upper RMS volt limit	Upper peak volt limit	Minimum clearance	Minimum creepage	Minimum clearance	Minimum creepage
12	17	0.19 (0.38)	0.40 (0.80)	0.19 (0.38)	0.40 (0.80)
30	43	0.28 (0.56)	0.55 (1.10)	0.28 (0.56)	0.50 (1.10)
60	85	0.38 (0.76)	0.72 (1.44)	0.38 (0.76)	0.72 (1.44)
100	141	0.62 (1.24)	1.12 (2.24)	0.62 (1.24)	1.12 (2.24)
125	177	0.62 (1.24)	1.12 (2.24)	1.60 (1.60)	1.60 (2.24)
130	184	0.62 (1.24)	1.12 (2.24)	2.40 (2.40)	2.40 (2.40)
250	354	1.15 (2.30)	1.95 (3.90)	2.40 (2.40)	2.40 (3.90)
380	540	1.75 (3.50)	2.80 (5.60)	9.50 (9.50)	12.7 (12.7)
500	710	2.40 (4.80)	3.70 (7.40)	9.50 (9 50)	12.7 (12.7)
600	850	3.60 (7.20)	5.60 (11.2)	9.50 (9.50)	12.7 (12.7)
750	1060	3.60 (7.20)	5.60 (11.2)	19.0 (9.0)	19.0 (19.0)
1000	1410	4.90 (9.80)	7.50 (15.0)	19.0 (19.0)	19.0 (19.0)
1250	1770	6.20 (12.4)	9.50 (19.0)	19.0 (19.0)	19.0 (19.0)
1500	2120	7.50 (15.0)	11.6 (23.2)	19.0 (19.0)	19.0 (23.2)
2000	2820	10.2 (20.4)	15.5 (31.0)	19.0 (20.4)	19.0 (31.0)
2500	3540	13.0 (26.0)	20.0 (40.0)	19.0 (26.0)	20.0 (40.0)
3000	4240	16.0 (32.0)	24.0 (48.0)	19 0 (32.0)	24.0 (48.0)

Note: If products are rated 60 Hz only with a secondary voltage of <100 VRMS 141 VPK/dc or if output is <200 VA, there are no specific spacing requirements and compliance is determined by the dielectric strength test.

Insulation class	Maximum °C
A (105)	100
E (For 50-Hz usage only)	115
B (130)	120
F (155, for 60-Hz usage only)	140
H (180, for 60-Hz usage only)	165

Temperature measurements are secured by the change of resistance method, where the transformer input voltage is supplied at 1.06 times the nominal rating and with the frequency at 50 Hz for transformers rated at 50 Hz, 50 to 60 Hz, or 50/60 Hz and at 60 Hz for transformers so rated.

Thermocouples may be employed for measurement purposes as a means of acceptance, when the temperature values shown in the table are reduced by 15°C. When noncompliance is determined by the thermocouple method, the change of resistance method may still be employed as a final determination of compliance.

11-2-7   UL and CSA Transformer Temperature Rating

The UL and CSA rate transformer temperature as a rise above ambient (25°C). Two methods are employed in temperature measurement, namely the thermocouple or the resistance method. The following table shows acceptable temperature rises.

	Maximum rise above ambient, °C	
Insulation class	Thermocouple method	Resistance method
105	65	75
130	85	95
155	110	120
180	125	135

REFERENCES

For in-depth and complete information on electrical safety standards, the reader is referred to the following original documents.

1. UL-478:   Electronic data processing units and systems.

2. UL-114:   Office appliances and business equipment—electric.

3. CSA-C22.2 No. 154-1975:   Data processing equipment.

4. CSA-C22.2 No. 143-1975:   Office machines.

5. IEC-380:   Office machines (safety of electrically energized office machines).

6. IEC-435:   Data processing equipment.

7. VDE-0730/Part 2P: Particular regulations for business machines.

8. VDE-0806/8.81:   Safety of electrically energized office equipment.

Index

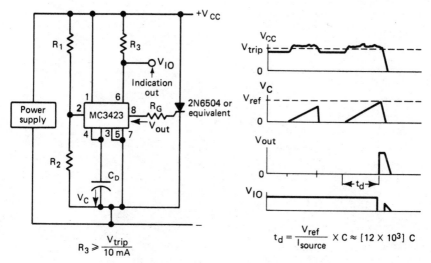

FIGURE 8-15 **Typical OVP application of the MC3423.** (*Courtesy of Motorola Semiconductor Products, Inc.*)

The earliest of these ICs was the MC3423, which has become an industry standard. A basic block diagram of this IC is given in Fig. 8-14. The diagram shows that the circuit consists of a stable 2.6-V reference, two comparators, and a high current output. The output is activated by a voltage greater than 2.6 V on pin 2, or by a high logic level on the remote activation, pin 5.

Figure 8-15 shows a typical application of the MC3423 in an OVP application. In the circuit, resistors R_1 and R_2 set the threshold trip voltage. The relationship between V_{trip} and R_1, R_2 is given by

$$V_{trip} = 2.6 \left(1 + \frac{R_1}{R_2}\right) \tag{8-11}$$

Keeping the value of R_2 below 10 kΩ for minimum drift is recommended.

The value of R_1 and R_2 may also be calculated by using the graph of Fig. 8-16. In the graph, $R_2 = 2.7$ kΩ, while R_1 may be directly calculated from the intersection of the trip voltage with the desired curve.

The MC3423 OVP circuit also has a programmable delay feature, which prevents false triggering when used in a noisy environment. In Fig. 8-15, capacitor C_D is connected from pins 3 and 4 to the negative rail to implement this function. The circuit operates as follows. When V_{CC} rises above the trip point set by R_1 and R_2, the internal current source begins charging the capacitor C_D connected to pins 3 and 4. If the overvoltage condition remains present long enough for the capacitor voltage V_{CD} to reach V_{ref}, the output is activated. If the overvoltage condition disappears before this occurs, the capacitor is discharged at a rate 10 times faster than the charging time,

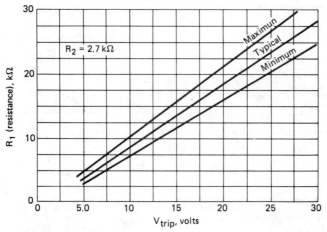

FIGURE 8-16 The threshold resistor values may be directly calculated from this graph, which plots R_1 vs. trip voltage for the MC3423 OVP circuit. (*Courtesy of Motorola Semiconductor Products, Inc.*)

resetting the timing feature. The value of the delay capacitor C_D may be found from Fig. 8-17.

A more elaborate OVP circuit, the MC3425, is in many respects similar to the MC3423, but the former one may also be programmed for undervoltage detection and also line loss monitoring. The block diagram of the MC3425 is shown in Fig. 8-18. Notice that this is a dual-channel circuit, with the overvoltage (OV) and undervoltage (UV) input comparators both referenced to an internal 2.5-V regulator. The UV input comparator has a feedback-activated, 12.5-μA current sink I_H, which is used for programming the hysteresis voltage V_H. The source resistance feeding this input R_H determines the amount of hysteresis voltage by

$$V_H = I_H R_H = (12.5 \times 10^{-6})R_H \qquad (8\text{-}12)$$

Separate delay pins 2 and 5 are provided for each channel to independently delay the drive and indicator output pins 1 and 6, respectively, thus providing greater input noise immunity. The two delay pins are essentially the outputs of the respective input comparators and provide a constant current source I_d of typically 200 μA when the noninverting input is greater than the inverting input level. A capacitor connected from these delay pins to ground will provide a predictable delay time t_d for the drive and indicator outputs. The delay pins are internally connected to the noninverting inputs of the OV and UV output comparators, which are referenced to the internal 2.5-V regulator. Therefore delay time t_d is based on the constant current source

About the Author

George C. Chryssis is founder and President of Intelco Corporation, a manufacturer of fiberoptic electronic test equipment.

Previously, he was co-founder and Vice President of Engineering of Power General Corporation, a successful power supply company. There he was responsible for all power supply design and development.

The author of numerous technical articles, he received B.S.E.E. and M.S.E.E. degrees from Northeastern University.